Proceedings in Engineering Mechanics

Research, Technology and Education

This book series publishes the results of meetings dealing with material properties in engineering and science. It covers a wide range of topics, from the fundamentals of materials mechanics and applications for various industries to aspects of scientific training and career development. The volumes in the series are based typically on primary research materials presented at conferences, workshops, and similar scientific meetings, and represent comprehensive scientific and technical studies.

More information about this series at http://www.springer.com/series/16733

Lucas F. M. da Silva · Robert D. Adams
Editors

6th International Conference on Adhesive Bonding 2021

Selected Contributions of AB 2021

Springer

Editors
Lucas F. M. da Silva
Department of Mechanical Engineering
Faculty of Engineering
University of Porto
Porto, Portugal

Robert D. Adams
Department of Mechanical Engineering
University of Bristol
Bristol, UK

ISSN 2731-0221 ISSN 2731-023X (electronic)
Proceedings in Engineering Mechanics
ISBN 978-3-030-87667-8 ISBN 978-3-030-87668-5 (eBook)
https://doi.org/10.1007/978-3-030-87668-5

This Springer imprint is published by the registered company Springer Nature Switzerland AG
The registered company address is: Gewerbestrasse 11, 6330 Cham, Switzerland

Preface

This volume is the first volume of the new book series *Proceedings in Engineering Mechanics—Research, Technology and Education.* This new series intends to publish the latest developments concerning fundamental research and applied technology of engineering mechanics. Education studies on science and technology are also the focus of this new series. The themes covered are very broad and have in common engineering mechanics. We want to go from fundamentals to applications and how these can be transferred to students.

This first volume of *Proceedings in Engineering Mechanics—Research, Technology and Education* contains selected papers presented at the 6th International Conference on Structural Adhesive Bonding (AB 2021), held in Porto (Portugal) during 8–9 July 2021 (www.fe.up.pt/ab2021). The goal of the conference was to provide a unique opportunity to exchange information, present the latest results as well as to discuss issues relevant to structural adhesive bonding research today. Approximately, 165 papers were presented by researchers from more than 27 countries.

In order to disseminate the work presented in AB 2021, selected papers were prepared which resulted in the present volume dedicated to adhesive bonding. A wide range of topics are covered resulting in 11 chapters dealing with the most recent research topics concerning adhesive bonding: adhesive formulation and properties (first four chapters), adhesion and surface treatments (three chapters), joint design (two chapters) and durability of structural adhesive joints (last two chapters). The book represents the latest trends in adhesive joining and serves as a reference volume for researchers and graduate students working with adhesive bonding.

The organizer and editor wish to thank all the authors for their participation and cooperation, which made this volume possible. Finally, I would like to thank the team of Springer-Verlag, especially Dr. Christoph Baumann, for the excellent cooperation during the preparation of this volume.

Porto, Portugal Prof. Lucas F. M. da Silva
lucas@fe.up.pt

Bristol, UK Prof. Robert D. Adams
August 2021

Contents

Open Assembly Time of Water-Borne Polyvinyl Acetate (PVAc) Wood Adhesives Affected by Various Factors

Peter Bliem, Hendrikus W. G. van Herwijnen, Robert Meissl, Walter Stadlbauer, and Johannes Konnerth

Abstract Open assembly time is one of the most important processing parameters of water-borne polyvinyl acetate (PVAc) dispersion adhesives. Exceeding the open assembly time may cause a loss of bonding properties. It is essential to have knowledge about possible factors influencing the open time. The present work evaluates different influencing factors on open assembly time of PVAc adhesives. At first, a simple and highly reproducible method was developed to determine the open time. Subsequently, by using this method, open times at varying relative humidities, room temperatures and application rates of PVAc adhesives were investigated on absorbent substrate. As a general result, this method is suitable as an alternative for determining the open assembly time. Furthermore, clear effects of some parameters investigated could be observed. While an increase in relative humidity and adhesive application rate led to a prolonged open assembly time, an increase in room temperature showed only little negative effect. However, if wood substrate and adhesive temperatures are simultaneously increased with the room temperature, a severe shortening of the open time could be observed.

P. Bliem (✉) · H. W. G. van Herwijnen
Wood K plus, Competence Centre for Wood Composites and Wood Chemistry, Altenberger Strasse 69, 4040 Linz, Austria
e-mail: p.bliem@wood-kplus.at

H. W. G. van Herwijnen
e-mail: e.herwijnen@wood-kplus.at

P. Bliem · H. W. G. van Herwijnen · J. Konnerth
Institute of Wood Technology and Renewable Materials, Department of Material Sciences and Process Engineering, University of Natural Resources and Life Sciences Vienna, Konrad-Lorenz-Strasse 24, 3430 Tulln an Der Donau, Austria
e-mail: johannes.konnerth@boku.ac.at

R. Meissl · W. Stadlbauer
H.B. Fuller Austria GesmbH, Bahnhofplatz 3, 4600 Wels, Austria
e-mail: robert.meissl@hbfuller.com

W. Stadlbauer
e-mail: walter.stadlbauer@hbfuller.com

L. F. M. da Silva and R. D. Adams (eds.), *6th International Conference on Adhesive Bonding 2021*, Proceedings in Engineering Mechanics,
https://doi.org/10.1007/978-3-030-87668-5_1

Keywords Polyvinyl acetate (PVAc) · Open assembly time · Adhesives for wood

1 Introduction

A wide variety of adhesive systems are available for bonding wood. The selection of the appropriate adhesive is typically based on the bonding requirements and/or the desired properties of the bonded joint or on process-related characteristics. A general distinction can be made between thermosetting and thermoplastic adhesive polymers. Polyvinyl acetate (PVAc) is an important representative of thermoplastic adhesives and, along with thermosetting urea formaldehyde (UF) adhesives, a commonly used adhesive system in the furniture industry (Dunky 2003).

The mainly physically setting of the water-borne dispersion adhesives are easy in handling, physiologically harmless due to the absence of formaldehyde emission or other reactants, and capable of forming an adhesive film at room temperature. The synthesis of PVAc is a chain growth reaction of vinyl acetate monomers dispersed in water in the presence of an emulsifier (usually polyvinyl alcohol) by free radical initiators (Frihart 2013; Jaffe et al. 1990; Pizzi 1989). The mean length of the resulting polymer chains is up to 3 μm and it forms many hydrogen bonds with the wood substrate due to the high content of acetate groups and the flexible characteristic of the polymer backbone (Frihart 2013; Zeppenfeld and Grunwald 2005). PVAc sets by water penetration into the wood substrate and/or by evaporation of the water into the ambient air. In the process of setting, the polymer particles gradually come closer to each other until they form a homogeneous adhesive layer (i.e. film formation) (Dunky 2003; Frihart 2013). The so-called setting speed can be influenced by two main groups according to Theiling (1972) as well as Rabiej and Brown (1985). The first group represents the properties of the adhesive itself, depending on the formulation chosen by the adhesive manufacturer. Important factors are the type of polymer system, protective colloid system, solid content, particle size of the dispersion, viscosity of the adhesive, type and amount of modifiers. On the other hand, the adhesive user as the second crucial group, has an influence on the setting speed of the PVAc adhesives. Several important parameters in this regard are the absorptivity of the substrate, the material moisture content and ambient humidity, adhesive application quantity, open assembly time, wood density, temperature (adhesive, wood substrate, ambient air), annual ring position, proportion of early and late wood and many more. Some of these factors can be considered and controlled when processing PVAc adhesives. Due to production processes, the open assembly time is often a critical factor in the processing of these adhesives, as a certain amount of time is required from applying the adhesive until joining the adherends (Sedliacik et al. 2014). The open assembly time is defined as the time starting from applying the adhesive to the adherend(s) until the assembly of the joint without loss of bonding strength (EN 923 2016). In other words, it is the time in which the applied adhesive contains enough water to sufficiently wet the opposite substrate material. Exceeding the open time will result in a dried adhesive layer and thus to poor bondings (Frihart 2013).

How quickly the water from the PVAc adhesive penetrates into the substrate or evaporates essentially depends on the adhesive, application, ambient and material conditions. In literature, some factors influencing the open time are mentioned, but there is little information available on how strong the influence actually is. While an increase in substrate and ambient temperature as well as the porosity of the substrate is related to a reduction in open time, a decrease in adhesive application quantity, substrate moisture content as well as relative humidity is attributed to a decrease in open time (Habenicht 2009).

Thickeners such as hydroxyethyl cellulose, methylcellulose, polyvinyl alcohol or starches will increase viscosity when added to an PVAc adhesive and release water slowly (Jaffe et al. 1990). As a result, they can prolong open time. Humectants as hygroscopic substances (i.e. glycerine, various glycols, sorbitol, sucrose, urea) in PVAc adhesives keep the surface wet and extend the open time, but because they hold and reabsorb water to a certain extent, they will soften the material after drying (Jaffe et al. 1990).

The maximum open times specified in the technical data sheets of the adhesive manufacturers are measured under defined laboratory conditions. These standardized conditions cannot be reached consistently in the processing companies, leading to undesirable shorter open times and thus to inadequately bonded joints.

A standardized method for measuring open time is defined in the standard EN 16556, which involves gluing beech lamellas together and measuring the tensile shear strength from the resulting lap joint samples (EN 16556 2014). Furthermore, it is also mentioned that the determination of the open time can be done by single lap joint samples according to EN 205 Annex A. Tankut et al. (2016), for instance, measured the effect of different nanoparticles and open times on bonding strength by means of lap joint samples. Another method, which is particularly used in practice, is to check the surface tackiness of the applied adhesive by means of a simple finger test. Since these methods are time and resource intensive or very user-dependent and therefore not really reproducible, an alternative method for determining open time of water-borne PVAc adhesives has been developed within the present study. With this method, conclusions can be drawn more quickly about possible factors influencing the open time. In addition, high reproducibility can be achieved with this method, since the solid wood with the typical high natural variability is replaced by more homogeneous materials. In this work, it is hypothesized that relative humidity, ambient air, material and adhesive temperature as well as adhesive application quantity are critical factors on the duration of open time. Therefore, these factors were systematically varied and open times were measured using the newly optimized method. The aim was to gain a better understanding of the change in open time by varying the various parameters and how strongly they affect the open time.

Table 1 Overview of the PVAc glue characteristics according to the technical and safety data sheets

Parameter	Standard D3 PVAc	Standard D4 PVAc
Solid content (%)	50	50
pH	3.0	3.5
Density at 20 °C (g/cm^3)	1.08	1.09
Viscosity (mPa * s)	13000 Brookfield HB, spindle 3, 20 rpm, + 20 °C	5500 Brookfield HB, spindle 2, 20 rpm, + 20 °C
Organic solvent content (%)	2.2	1.8

2 Materials and Methods

In the first step, an alternative method for determining the open time of PVAc adhesives was developed. This method should be reproducible, easy in handling, with a maximum accuracy and provide prompt results. Subsequently, the open times at varying relative humidities, room temperatures and application rates of PVAc adhesives were investigated and analysed.

2.1 *Adhesive System Characteristics*

Two different industrial polyvinyl acetate glues were used in this study (H.B. Fuller Austria GesmbH, Wels, Austria). The main characteristics for both glues are listed in Table 1. Standard D3 PVAc fulfils the durability class D3 according to EN 204 (EN 204 2016) and is commonly used for veneering and surface/assembly gluing. The standard D4 PVAc is a one component PVAc glue offering improved water and heat resistance used for the production of window frames and doors (durability class D4).

2.2 *Experimental Designs*

In the first part of the present study, the influences of different environmental conditions, adhesive application quantities, and PVAc glue types on open time were investigated and Table 2 shows the experimental designs. The relative humidity was systematically changed, starting from 35 to 50% up to 65%. The ambient temperature of the testing room had two levels, namely 23 or 33 °C. The MDF (medium density fibre) board and adhesive temperature were always set to 23 °C. Therefore, MDF boards were first stored in standard climate and then tempered to the desired temperature in

Table 2 Experimental design with various possible influencing parameters for investigating open time

parameter			
relative humidity (%)	35 – 50 – 65		
ambient temperature (°C)	23 – 33		
MDF board and adhesive temperature (°C)	23		23 or 33 (corresponding to ambient temperature)
glue application quantity (g/m²)	105 – 135		
PVAc glue	standard D3 PVAc	standard D4 PVAc	standard D3 PVAc

a climate chamber shortly before use. PVAc glues, in contrast, were stored airtight inside the climate chamber at 23 °C.

In the second part of the work, the MDF board and D3 PVAc glue temperatures were also increased to 33 °C in case of an increased testing room temperature. The procedure for storing the boards and the adhesive was identical as described above, but now at 33 °C. Subsequently, the boards and the adhesive temperature corresponded to the ambient temperature of the testing room. This experiment was not conducted with the D4 PVAc in order to prevent premature reaction of the adhesive at correspondingly higher temperature.

A manual film applicator with gap clearances of 150 and 200 μm and a width of 60 mm was used for adhesive application (mtv messtechnik oHG, Erftstadt, Germany). Accordingly, these gaps resulted in adhesive quantities of 105 g/m^2 and 135 g/m^2, respectively.

2.3 *Influence of Temperature on Viscosity*

Since part of the study involved tempering the D3 PVAc glue to 33 °C, the viscosity of this glue was measured depending on the temperature. D4 PVAc glue was not tempered to 33 °C in the experiments, but a temperature ramp was also measured as additional information. Both temperature ramps were performed at testing temperatures between 22 and 34 °C at a constant shear rate of 10 1/s using a rheometer (MCR 302, Anton Paar GmbH, Austria) equipped with a cone-plate measuring system (CP 50–1, Anton Paar GmbH, Austria). Gap size was set to 0.101 mm resulting in a test volume of 0.57 ml.

2.4 Procedure for Determining the Open Time

The entire experiment was performed in a climate-controlled room exhibit a volume of 30 m^3 at the technical centre of Wood K plus, Area Wood Materials Technologies (Tulln/Donau, Austria). Target air humidity and room temperature were set according to the experimental designs and monitored twice using a Sensirion digital sensor (SHTW2, Staefa, Switzerland) and a data logger (Lascar EL-USB-2+, Wiltshire, United Kingdom). The accepted fluctuation ranges of the measured climate conditions were ±4% for relative humidity and ±2 °C for room temperature. During the individual measurements, the actual values were measured and used for the data evaluation.

Figure 1 shows the schematic procedure for measuring open time and is a development of an unpublished laboratory method used in the adhesive producing industry. In step 1, the tempered PVAc glue was applied on a likewise tempered MDF board (Kaindl MDF Standard E1E05, M.Kaindl OG, Wals, Austria) using a manual film applicator. Dimensions of the MDF board were thereby 600 × 200 × 19 mm (length × width × thickness) with an average board density of 700 kg/m^3. All MDF boards for the individual experiments performed were previously stored at standard climate conditions (20 °C and 65% relative humidity) until constant mass was reached, resulting in an average moisture content of 8.6%. MDF board was tempered to the desired temperature (i.e. 23 or 33 °C) inside a separate climate chamber prior to adhesive application. The relative humidity inside the climate chamber has been slightly increased to keep the moisture content of the MDF board constant and prevent undesired drying or influence of variable substrate moisture content. Therefore, the relative humidities were set to 66% (at 23 °C) and 68% (at 33 °C) respectively. In addition, the weight of the MDF boards were measured before and after storage to determine any possible changes in the moisture content. The use of MDF instead of another wood-based material or solid wood seemed to be the best solution to avoid undesired influences of less homogeneous materials (i.e. density differences, fibre deviations, different permeability of the surface, etc.). The actual amount of glue applied was determined gravimetrically by means of a scale with an accuracy of 0.01 g. A few minutes were then waited so that the previously applied glue could dry off a little. Then a simple strip of copy paper, as described below, was placed on the glue film and weighted down with a piece of MDF (out of the same material as the MDF board) and a weight of 1 kg (step 2). The dimensions of the paper strip and the MDF piece were 120 × 45 mm (length × width), with an overlapping area of these on the MDF board of 90 × 45 mm. This resulted in a specific pressure of 2421 Pa. Copy paper with a grammage of 80 g/m^2 (Nautilus® Classic, Mondi AG, Austria) was used for various tests. Preliminary experiments showed influences of paper direction and paper side used on the later determined fractured surface. To reduce variability of results in the open time, the paper strips were always glued to the glue film with the same direction and side. In step 3, additional paper strips were glued on the glue film every minute, if longer open times were expected. For shorter open times, especially when using higher MDF board and glue temperatures

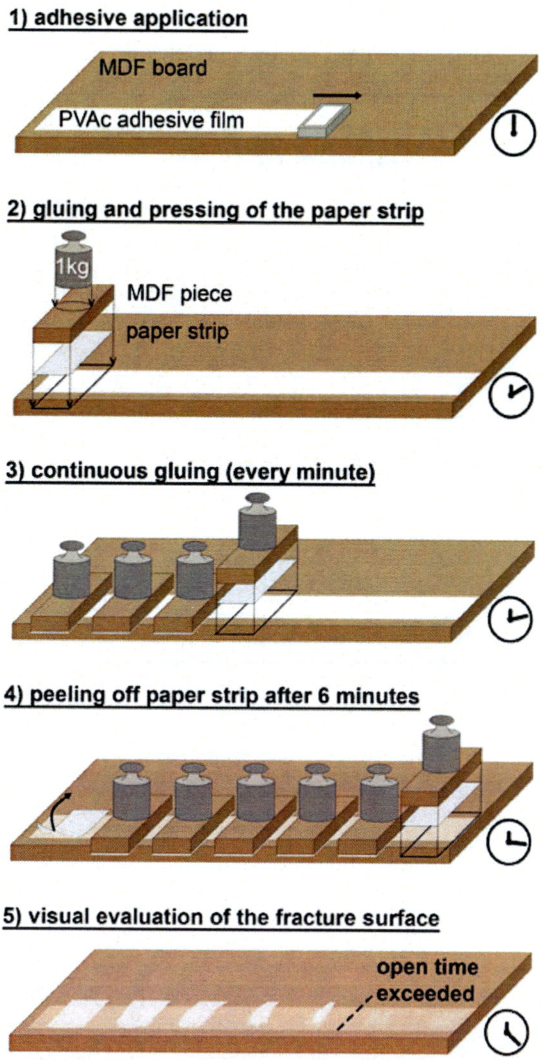

Fig. 1 Schematic procedure for measuring the open time of PVAc glues

in combination with a low relative humidity, every half minute was tested. Each paper strip was pressed for 6 min before being manually peeled off (step 4). The visual evaluation of the fractured surface was examined after all strips were peeled off. As soon as no larger fragments of the paper remained on the glue film, the open time (time between glue application and assembling of the joint) was considered to be exceeded (step 5). Compared to the standard EN 16556, the proposed method does not measure strength properties of the joint, but only whether the surface of the adhesive still contains enough moisture to wet the paper sufficiently. For each experimental testing point in Table 2, five repetitions were performed to determine the open time.

2.5 *Statistical Analyses*

From all 5 repetitions, the arithmetic mean and the corresponding standard deviation were calculated. Furthermore, regression models were calculated to identify significant factors influencing the open times depending on the experiments carried out. An F-test with a significance level of 5% was used to select statistically significant variables and their interactions.

3 Results and Discussion

As a first result, it can be stated that the presented method is well suited for measuring the open assembly time of PVAc adhesives. Compared to other methods, reproducible results can be achieved without the need for large amounts of material and time. Furthermore, the method can also be used to investigate the effects of modifications of the adhesive itself on the open time.

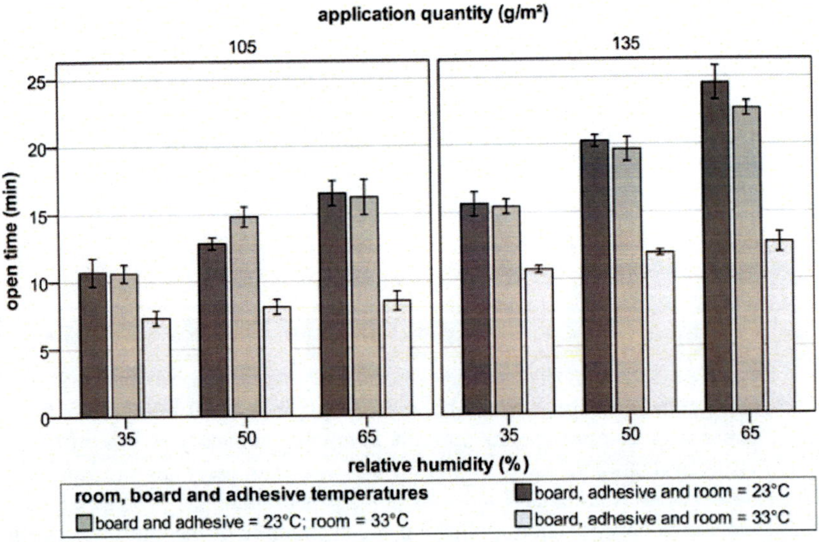

Fig. 2 Open times of D3 PVAc glues at various relative humidities, temperatures and application quantities

The measured open times of the D3 PVAc glue are illustrated in Fig. 2. Each individual bar represents the mean value of five repetitions performed, with the standard deviation displayed as an error bar. A comparison of the two different application quantities shows a clear difference in open time. The more glue is applied, the longer the maximum open time, which is also mentioned in literature (Dunky and Niemz 2002). Interestingly, similar open assembly time values can be achieved with a lower application quantity of 105 g/m^2 at high relative humidity (i.e. 65%) than with a higher application rate of 135 g/m^2 and a simultaneous low relative humidity (i.e. 35%). As a result, the amount of adhesive can be significantly reduced without reducing the open time if air humidity near the adhesive application would be kept high enough. This could lead to distinct savings in glue quantities for PVAc-processing companies, where open assembly time is the limiting factor in the process. However, the relative humidity exhibits a positive influence on the open time for both application quantities. This behaviour is also mentioned by Habenicht (2009). The author further stated that a higher moisture content of the substrate material would cause a prolonged open time, which was not investigated in this study.

The temperatures of the room, MDF board and adhesive influence the open time very differently. Comparing the results in Fig. 2, the measured open time behaves quite similar for each individual relative humidity and PVAc application quantity, when MDF board and glue temperature are always kept constant at 23 °C. It can be concluded, that there is no considerable change in the open time even if the ambient temperature rises to 33 °C. However, increasing the MDF board and glue temperature to 33 °C leads to a significant reduction of open time as illustrated in the results in Fig. 2. Habenicht (2009) already stated a reduction of the open time with a simultaneous higher temperature of the adherend or ambient air.

Based on the results in Fig. 2, it can be observed that the relative extension of the open time exhibits similar relative changes of open time for the two different glue application quantities. An increase in relative humidity from 35 to 50% seems to be more effective than the increase from 50 to 65% as can be seen by the higher relative changes of the open time. The total relative increase of open time from 35 to 65% relative humidity resulted in prolongations between 47 and 57% for the two glue application quantities and room temperatures investigated. Using a room, MDF board and adhesive temperature of 33 °C, a considerably smaller relative increase in open time compared to a temperature of 23 °C could be observed. Whereas a change in relative humidity from 35 to 65% at 23 °C room temperature resulted in a relative increase in open time between 54 and 57%, values of only 16 and 19% could be achieved at a room, MDF board and adhesive temperature of 33 °C. A possible reason for this could be the lower viscosity due to the increased adhesive temperature as shown in Fig. 3 and hence a faster penetration behaviour.

Figure 4 shows the open times of the investigated D4 PVAc glue at different relative humidities, room temperatures and applied glue quantities. The MDF board and glue temperature were always kept at 23 °C to prevent a premature reaction of the D4 PVAc glue. In general, the mean open times are insignificantly lower compared to the results of the D3 PVAc in Fig. 2. A possible reason for this might be the considerably lower viscosity of the D4 PVAc glue as listed in Table 1 and also measured in Fig. 3,

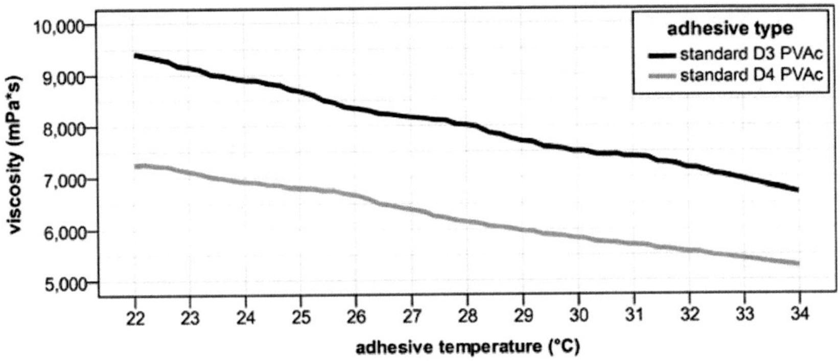

Fig. 3 Viscosity—temperature relationship of the two PVAc glues investigated at a constant shear rate of 10 1/s

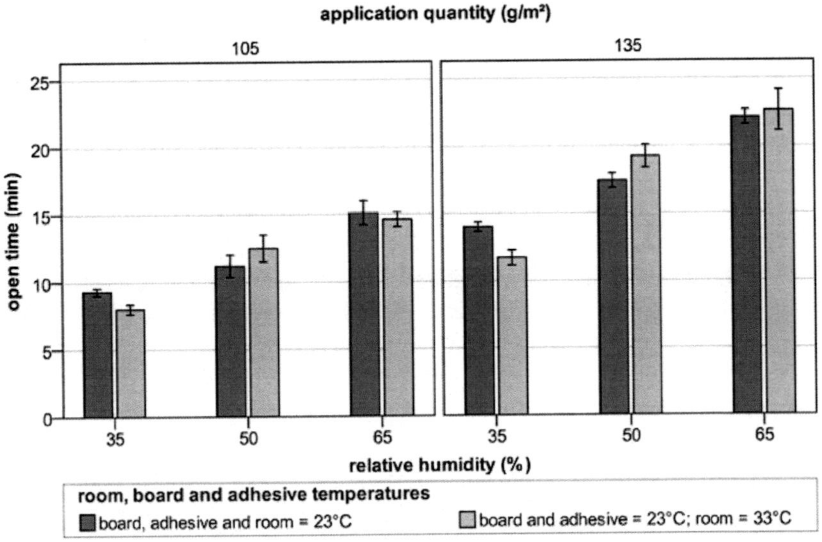

Fig. 4 Open times of D4 PVAc glues at various relative humidities, temperatures and application quantities

which could led to a slightly accelerated penetration of the glue into the MDF board and thus to a faster drying of the applied glue film. For both application quantities, i.e. 105 and 135 g/m^2, the positive influence of increasing relative humidity can be clearly seen. The total relative increase in open time, i.e. from 35 to 65% relative humidity, was 83–93% for 33 °C room temperature, whereas slightly lower value of 58–62% could be achieved at 23 °C room temperature. Based on the results, it seems that the influence of the change in relative humidity is more crucial at higher ambient temperature than at lower room temperature. Furthermore, D4 PVAc glue in Fig. 4 showed the same behaviour as the D3 PVAc glue in Fig. 2, where similar open times could be reached with a lower applied glue quantity (i.e. 105 g/m^2) at simultaneous high relative humidity of 65% compared to a higher application quantity of 135 g/m^2 at 35% relative humidity. This, in turn, could lead to savings in PVAc glue quantities without deteriorating the bonding properties.

Controlling the relative humidity is definitely a possibility and is more or less done in industry, but conditioning the substrate material together with the adhesive is rather difficult to implement. On the other hand, the manufacturer's processing guidelines must be strictly followed, otherwise poor bonding properties may result. The adhesive temperature and especially the substrate material usually have the same temperature as the ambient temperature in the production hall. At elevated ambient and substrate temperatures, cooling the adhesive to 23 °C alone would be easier to implement. Whether this could also lead to an analogous effect was not investigated in this work, but would be conceivable for further research.

As a further statistical analysis, three regression models have been calculated with adjusted overall coefficients of determination (R^2) of 0.94, 0.96 and 0.96, indicating a rather good estimation of the open time described by the corresponding influencing model parameters. The scaled regression coefficients for the first model, where MDF board and D3 PVAc adhesive temperatures were always kept at 23 °C, are listed in Table 3. The coefficients show the relative impact of the parameters as linear as well as two-factorial interaction terms on the open time of PVAc glue. While relative humidity (A) and the glue application quantity (C) as linear terms showed a distinct positive influence on the response value (i.e. open time), room temperature was not significant for describing the model. This positive influence can be clearly seen in the 3D model plot in Fig. 5a, where open time increased rapidly due to the increase in relative humidity (A) or glue application rate (C). The corresponding interaction (AC) of relative humidity (A) and application quantity (C) affected the open time slightly positive. Furthermore, the interaction (AB) of relative humidity (A) and room temperature (B) had a negative influence on the open time.

For the second model, where the MDF board and D3 PVAc adhesive temperature were dependent on the room temperature used (i.e. 23 or 33 °C), the two factors

Table 3 Influencing factors and their interactions of processing parameters, where MDF board and D3 PVAc adhesive temperatures were always kept at 23 °C, to predict the open time using multiple linear regression (level of significance = 0.05; n.s. = not significant)

Model parameters		Scaled regression coefficients	p-value
A	Relative humidity (%)	+3.25	<0.0001
B	Room temperature (°C)	n.s	–
C	Glue application quantity (g/m^2)	+2.93	<0.0001
AB	Interaction A × B	−0.79	0.0031
AC	Interaction A × C	+0.55	0.0047
BC	Interaction B × C	n.s	–

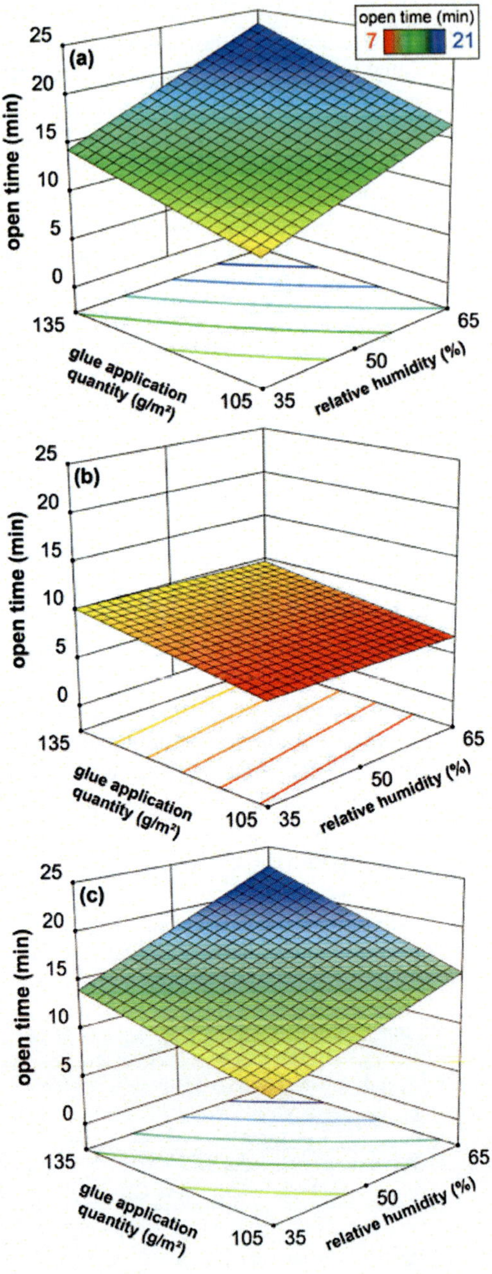

Fig. 5 **a** 3D model plots of the results from the D3 PVAc glue at a room, MDF board and adhesive temperature of 23 °C, **b** from the D3 PVAc glue at a room, MDF board and adhesive temperature of 33 °C and **c** from the D4 PVAc glue at a room, MDF board and adhesive temperature of 23 °C

Table 4 Influencing factors and their interactions of processing parameters, where MDF board and D3 PVAc adhesive temperatures were corresponding to ambient temperature, to predict the open time using multiple linear regression (level of significance = 0.05; n.s. = not significant)

Model parameters		Scaled regression coefficients	p-value
A	Relative humidity (%)	+1.67	<0.0001
B	Room, MDF board and adhesive temperature (°C)	−4.01	<0.0001
C	Glue application quantity (g/m²)	+2.22	<0.0001
AB	Interaction A × B	−1.90	<0.0001
AC	Interaction A × C	n.s	–
BC	Interaction B × C	−0.87	<0.0001

relative humidity (A) and glue application quantity (C) again showed a significant positive impact on the open time (Table 4). Furthermore, temperature of the room, MDF board and adhesive (B) as a linear factor was also significant to predict the open time using multiple linear regression model. Nevertheless, the temperature of the room, MDF board and adhesive showed the strongest impact on the open time, but negatively, so that an increase in the temperature led to a considerable reduction in open time. The two-factor interactions (AB) and (BC) also had a negative impact on the open time. Figure 5b shows the 3D model plot when room, MDF board and adhesive temperature were kept at 33 °C. By direct comparison of the 3D model plots between Fig. 5a, where all temperatures were set to 23 °C and Fig. 5b, the effect of an increased MDF board and adhesive temperature can be clarified. This could have an important negative consequence on the bonding properties for PVAc processing companies, particularly in case the open time has been exceeded due to the use of too warm adhesives or wood substrates to be bonded.

As third model, Table 5 shows the influencing parameters using the D4 PVAc glue. Figure 5c illustrates the 3D model plot at a room temperature of 23°. The relative

Table 5 Influencing factors and their interactions of processing parameters, where MDF board and D4 PVAc adhesive temperatures were always kept at 23 °C, to predict the open time using multiple linear regression (level of significance = 0.05; n.s. = not significant)

Model parameters		Scaled regression coefficients	p-value
A	Relative humidity (%)	+3.75	<0.0001
B	Room temperature (°C)	−0.75	<0.0001
C	Glue application quantity (g/m²)	+2.84	<0.0001
AB	Interaction A × B	n.s	–
AC	Interaction A × C	+0.62	<0.0001
BC	Interaction B × C	n.s	–

humidity (A) had the strongest positive influence on the open time followed by the glue application quantity (C) and the two-factor interaction of (AC). Only the room temperature (B) showed a slightly negative effect on the open time of the D4 PVAc glue.

4 Conclusion

In summary, it can be said that by using the described method, the open assembly time of PVAc adhesives can be checked easily, efficiently and with high reproducibility.

Selected parameters were examined which could be of importance for operational practice. For the processing of PVAc, the open time is an important parameter that must be carefully considered in order to avoid any loss of bonding performance. The relative humidity turned out to be a significant influencing factor on the open time. Similarly, a higher glue application quantity leads to a prolonged open time, but it is a decisive cost factor as well and a too thick bond line should be avoided. An increased room temperature, as occurs in non-air-conditioned production halls during the summer months, has little effect on the open time of PVAc. This is only true if the substrate material together with the adhesive are kept cool within the proposed processing parameters. If this is not the case, increasing the wood substrate and adhesive temperature will lead to a significant reduction in open time. By specifically controlling the parameters in this study, the open time should be reasonably predictable. If longer open times are required for production reasons, the addition of additives (Jaffe et al. 1990; Zeppenfeld and Grunwald 2005; Tankut et al. 2016; Sekisui Specialty Chemicals … (https://www.sekisui-sc.com/wp-content/uploads/2020/10/SelvolPVOH_AdhesivePolymer_EN.pdf); Ferguson 2000; Kremer Pigmente … (https://www.kremer-pigmente.com/elements/resources/products/files/67700-67760e.pdf); Minelga and Norvydas 2005; Grünewälder et al. 1998) or the foaming of PVAc adhesives (Sedliacik et al. 2014) could also be considered.

References

Dunky, M.: Chapter 47 adhesives in the wood industry. In: Pizzi, A., Mittal, K.L. (eds.) Handbook of Adhesive Technology, p. 672. CRC Press, Boca Raton (2003)

Dunky, M., Niemz, P.: Holzwerkstoffe und Leime: Technologie und Einflussfaktoren (in German). Springer, Berlin Heidelberg (2002)

EN 16556: Determination of the maximum open time for thermoplastic wood adhesives for non-structural applications (2014)

EN 204: Classification of thermoplastic wood adhesives for non-structural applications (2016)

EN 923: Adhesives—Terms and definitions (2016)

Ferguson, C.J.: Core-shell polymers styrene and acetate use as wood adhesives (2000)

Frihart, C.R.: Chapter 9 wood adhesion and adhesives. In: Rowell, R.M. (ed.) Handbook of Wood Chemistry and Wood Composites, p. 687. CRC Press, Boca Raton (2013)

Grünewälder, B., Parent, B., Dziallas, M., Haller, W., Wegner, J.: Aqueous dispersion adhesive based on polyvinylacetate (1998)

Habenicht, G.: Applied Adhesive Bonding: A Practical Guide for Flawless Results. WILEY-VCH, Weinheim (2009)

Jaffe, H.L., Rosenblum, F.M., Daniels, W.: Polyvinyl acetate emulsions for adhesives. In: Skeist, I. (ed.) Handbook of Adhesives, p. 800. Chapman & Hall, New York (1990)

Kremer Pigmente GmbH & Co. KG: Product information 67700–67760 Polyvinyl Alcohol. https://www.kremer-pigmente.com/elements/resources/products/files/67700-67760e.pdf

Minelga, D., Norvydas, V.: Properties of halogensilane modified poly(vinyl acetate) dispersion. Mater. Sci. **11**, 146–149 (2005)

Pizzi, A.: Wood Adhesives: Chemistry and Technology. CRC Press, London (1989)

Rabiej, R.J., Brown, P.T.: Status and needs. In: Proceedings Wood Adhesives. pp. 49–74. Madison (1985)

Sedliacik, J., Schmidtova, J., Smidriakova, M.: Foaming of polyvinylacetate adhesives—improved technologic parameters for a solid wood gluing process. Ann. Warsaw Univ. Life Sci.—SGGW. For. Wood Technol. **87**, 224–228 (2014)

Sekisui Specialty Chemicals Europe S.L: A versatile polymer for adhesive applications. https://www.sekisui-sc.com/wp-content/uploads/2020/10/SelvolPVOH_AdhesivePolymer_EN.pdf

Tankut, N., Bardak, T., Sozen, E., Tankut, A.N.: The effect of different nanoparticles and open time on bonding strength of poly (vinyl acetate) adhesive. Measurement **81**, 80–84 (2016). https://doi.org/10.1016/j.measurement.2015.12.003

Theiling, E.A.: Einflüsse auf die Abbindegeschwindigkeit von Dispersionsklebstoffen (in German). Adhäsion. 18 (1972)

Zeppenfeld, G., Grunwald, D.: Klebstoffe in der Holz- und Möbelindustrie (in German). DRW-Verlag, Leinfelden- Echterdingen (2005)

Engineering Technology and Preparation of Coating Systems Made with Epoxy Resin-Based Bonding Agents Modified with the Addition of Granite Powder Sourced from Quarry Wastes

Łukasz Kampa and Łukasz Sadowski

Abstract This article describes the engineering technology and preparation of coating systems made of epoxy resin-based bonding agents modified with the addition of granite powder sourced from quarry wastes. Particularly, it presents the effect of adding granite powder sourced from quarry wastes to the epoxy resin bonding agent on its pull-off strength. The analyzed bonding agent was laid on the specially prepared substrate made of concrete. The bonding agent was modified with waste granite powder in the mass proportions ranged from 10 to 60%. The obtained results were compared with the reference coating systems made without the addition of powder. The best results were observed for the bonding agent with the addition of 10 and 20% of waste granite powder compared to the reference coating systems. The addition of an amount higher than 20% of the waste granite powder resulted in a decrease or a slight increase in the pull-off strength compared to the reference coating systems. Additionally, the bonding agent becomes completely viscous when added above 50%. Economic analysis shows that such epoxy resin-based bonding agents allows for considerable savings. In the case of the epoxy resin-based bonding with the highest pull-off strength, due to the low cost of granite powder, it allows savings of up to 20%.

Keywords Epoxy resin · Bonding agent · Pull-off strength · Granite powder · Quarry waste

Ł. Kampa (✉) · Ł. Sadowski
Wroclaw University of Science and Technology, Wybrzeże Wyspiańskiego 27, 50-370 Wroclaw, Poland
e-mail: lukasz.kampa@pwr.edu.pl

Ł. Sadowski
e-mail: lukasz.sadowski@pwr.edu.pl

L. F. M. da Silva and R. D. Adams (eds.), *6th International Conference on Adhesive Bonding 2021*, Proceedings in Engineering Mechanics,
https://doi.org/10.1007/978-3-030-87668-5_2

1 Introduction

Epoxy resin-based bonding agents are popular in strengthening concrete before the application of protective coatings. On the other hand, waste granite powder, which is waste from the mining industry, has little use, or rather it is an attempt at its disposal. It is used as ballast for concrete or paving stones, or even as fertilizer. Recently, powders sourced from quarry wastes are quite popular additives in cementitious composites. The study (Sadeka et al. 2016) presents the effect of adding marble powder, granite powder, and mixed powder on self-compacting concrete. Granite powder turned out to be better, but both resulted in improved mechanical properties of the product, including tensile and flexural strength. On the other side, Li et al. (2019) added granite powder to reduce the amount of added cement and at the same time increase the strength of the mortar. They pointed out that this method allows for the removal of waste (granite powder) and the carbon footprint. This technology, as the authors point out, is better than a method of replacing cement to utilize waste and reduce cement content. It also improves the strength and microstructure of the tested mortar. Chajec (2021) checked the effect of adding granite powder and fly ash to cement mortars. Both materials allow to reduce the amount of cement in the mortar, which at the same time contributes to the reduction of CO_2 emissions generated in the production of cement. The use of granite powder significantly reduces the cost of production. Rizwan et al. (2021) tested the effect of coating concrete slabs with polychonic spray. Regardless of the specific properties of the substrate, it allows the stiffness, peak load, plastic load, and plasticity to increase. An equally important area in which cement is used are adhesives for gluing walls and floor tiles. Francke and Piekarczuk (2020) carried out a test of adhesion between waterproof coatings and terrace tiles under operational loads. They found that water has the greatest effect on reducing the tensile strength of the substrate of liquid products under the tiles, the adhesion of waterproof coatings to the substrate is greater than the adhesion between the waterproof coating and the tile adhesive layer, regardless of the substrate.

There are also studies describing the influence of various types of modification methods used to improve the properties of epoxy resins in coating systems. The properties of such coatings are shown in Table 1.

The studies presented above show that by using various types of materials, sometimes even waste, it is possible to improve certain properties of the coatings. However, there are no studies showing the effect of adding waste granite powder to epoxy resin-based bonding agents. Considering the above, the main aim of this article is to describe the engineering technology and preparation of coating systems made of epoxy resin-based bonding agents modified with the addition of waste granite powder sourced from quarry wastes. Particularly, it presents the effect of adding waste granite powder sourced from quarry wastes to the epoxy resin bonding agent on its pull-off strength. The analyzed bonding agent was laid on the specially prepared substrate made of concrete. The bonding agent was modified with waste granite powder in mass proportions ranging from 10 to 60%. The obtained results were compared with the reference coating systems made without the addition of powder.

Table 1 Selected examples of epoxy coating modifications and their effects

Authors and references	Modification method	Effect
Othman et al. (2020)	Highly dispersed graphene oxide–zinc oxide nanohybrids	Increased corrosion resistance of the coating
Morshed et al. (2020)	Nanosilica and multiwall carbon nanotubes	Improves the bond between polymer composites
Pourhashema et al. (2017)	Hybrid SiO_2-graphene oxide	Improves corrosion resistance, adhesion, and properties in contact with water
Chowaniec et al. (2020)	Glass powder	Improving the adhesion of the epoxy coating, without any mechanical treatment of the substrate
Do and Soh (2003)	Polymers	Improving the adhesion of the epoxy coating, without any mechanical treatment of the substrate
Ahn (2003)	Monomers	Improving the adhesion of the epoxy coating, without any mechanical treatment of the substrate
Atta et al. (2017)	Graphene Oxide	Lower water absorption of the coating
Chi et al. (2020)	Resin combined with solvent by Diels-Adler reaction	Better anticorrosion protection, increased durability, reduced viscosity, increased adhesion and mechanical strength
Kampa et al. (2020)	Polypropylene fibers	Greater pull-off strength
Chowaniec and Ostrowski (2018)	Glass powder	Increase in pull-off strength
Krzywiński and Sadowski (2019)	Modification of the substrate by texturing	Increase in pull-off strength
Kim et al. (2019)	Aminosilanams	Increase in pull-off strength
Patil et al. (2019)	Synthesis of biobased epoxy resin from gallic acid with various epoxy equivalent weights	Improvement of mechanical, chemical, and thermal properties
Sun et al. (2021)	Catechol	Increase in pull-off strength

2 Materials and Methods

2.1 Concrete Substrate

The substrate was made of concrete class C30/37, which was ordered from a specialized company producing concrete mixes. The degree of watertightness of this concrete was W8, consistency class S3, and a maximum grain size of 16 mm. The air content was 2.5 ± 1%. Concrete was stored for 28 days at 20 °C.

The cement used is CEM III/A 42.5 N-LH/HSR/NA. Fly ash and OMEGA 132 admixtures were also used. The W/C ratio was 0.48. Aggregate with the 0/2 dimension was 39% with a moisture content of 5% and a density of 2650 kg/m^3, 2/8 was 24% with a moisture content of 3%, and a density of 2570 kg/m^3, and the 8/32 aggregate content was 37% with a moisture content of 2% and a density of 2570 kg/m^3.

2.2 *Waste Granite Powder*

The granite powder used in the epoxy coating was wasted from the extraction of granite aggregate. The analysis of the chemical composition shows that the major part is SiO_2—53.63%. The Al_2O_3 content was 19.9%, CaO 8.3%, MgO 5.54%, FeO 3.46%, NaO 3.46%, K_2O 3.29% and Fe_2O_3 2.42%. The distribution of individual components is shown in Fig. 1a.

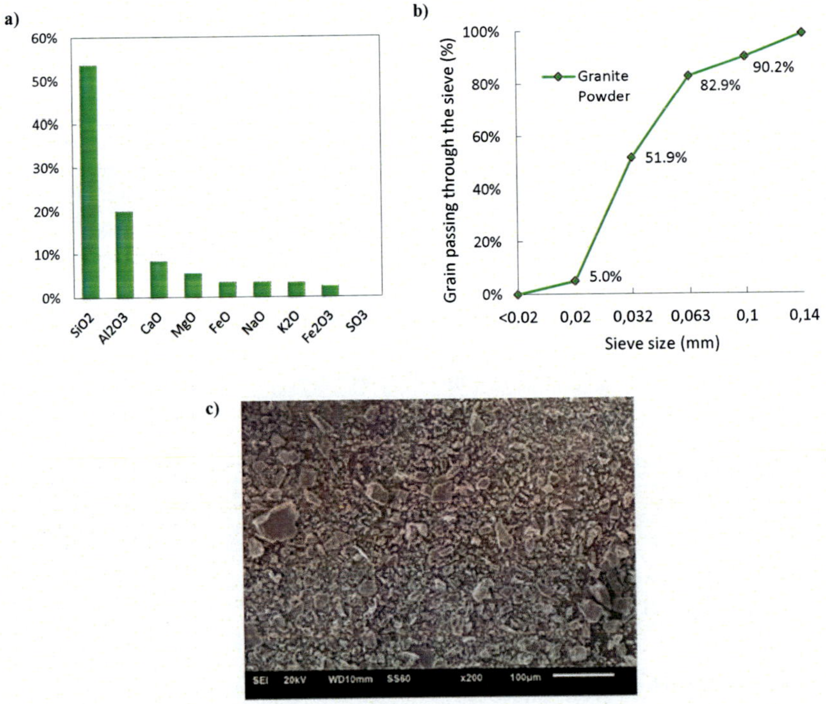

Fig. 1 Basic properties of waste granite powder: **a** chemical composition, **b** graining curve, **c** morphology

To control the grain thickness, an analysis was carried out on the control sieves. The fraction up to 0.02 mm accounted for 5%, 0.02–0.032 mm was the largest proportion—46.9%. 31% was the fraction 0.032–0.063 mm, 7.3% the fraction 0.063–0.1 mm, 8.7% the fraction 0.1–0.14 mm, and above 0.14 mm—1.1%. The graph of the graining curve is shown in Fig. 1b.

2.3 Epoxy Resin

The epoxy resin Meteor Primer (Si-Tech Sp. Z oo, Dobra 9, 05-306, Jakubów) was the modified base layer in this study. It was characterized by a density of 1–1.2 g/cm^3, a viscosity of 400–600 MPa · s, pot life (at 20 °C) 20–30 min and full hardening after 24 h. The resin is the result of the reaction of bisphenol A with epichlorohydrin (epoxy resin—average molecular weight ≤ 700). Its basic parameters are: reaction to fire—Bfl-s1, release of corrosive substance—SR, abrasion resistance ≤ AR1, impact strength ≥ IR4, and pull-off ≥ B1.5 (manufacturer's data). Only the primer was used in the test. A typical epoxy floor still contains a protective topcoat, but in this study it was not used due to the fact that both of these layers burn much stronger than the substrate—epoxy coating. The modification procedure is described in the next section.

2.4 Preparation of Coating Systems

The concrete substrate made earlier was carefully sanded with fine sandpaper. After that, the surface was vacuumed of all impurities and degreased with acetone. The individual research fields were separated with a bituminous roofing sealant. Each area was 20 × 20 cm. Then the resin was prepared. The resin consisted of two components: A and B in the ratio 100:33. In this study, it was decided to test the addition of 10, 20, 30, 40, 50, and 60% by weight. waste granite powder. The appropriate amount of component A was poured into a tared plastic cup placed on a balance. Waste granite powder was then added while weighing it. After the two elements were mixed, the appropriate amount of component B was added. The whole mixture was mixed for 2 min. The more powder was added, the blacker the mixture turned, even though the powder was gray. For control purposes, a costing system was also made without the addition of powder. The mixture was poured on individual research fields, creating a coating with a thickness of about 3 mm. The mass flowed well over the surface, even with a larger amount of powder. It was vented with a brush with hard bristles. Figure 2 shows the cross-section through the tested coating.

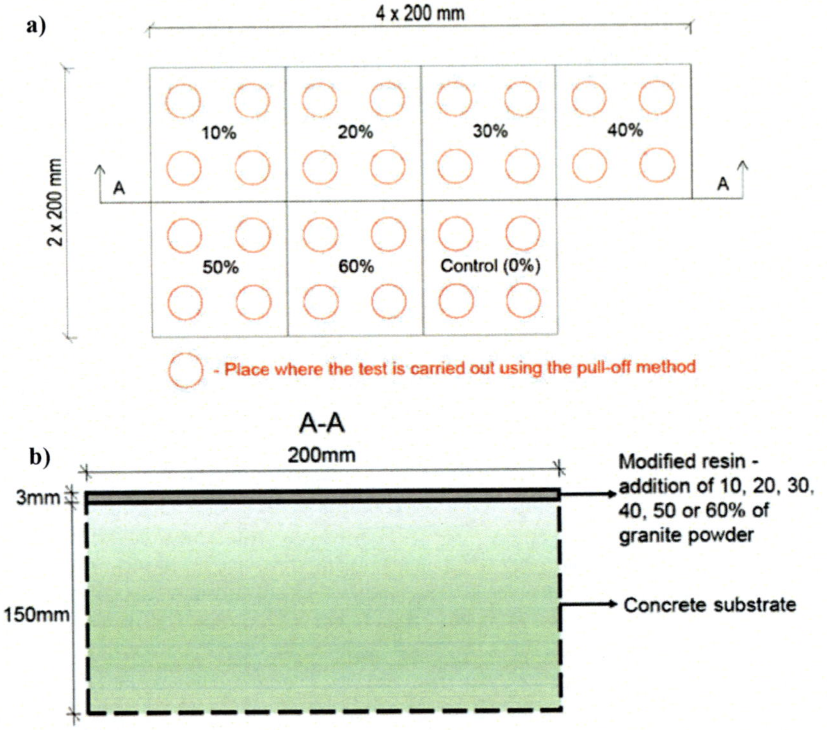

Fig. 2 Tested coating: **a** division into individual test fields (the value in the field determines the amount of waste granite powder added), **b** cross-section through the coating

2.5 *Pull-Off Strength Test of Epoxy Resin-Based Bonding Agents*

Two weeks after the invention of the resin coating, preparations for the pull-off strength test were started. The coating was sanded with fine sandpaper to better adhere the steel disc, which will connect the coating to an automatic device for testing the pull-off strength. Then, the drillings with a diameter of 50 mm and a depth of 5 mm were made. Four boreholes were drilled on each of the additions, which were drilled with a diamond hole saw. In the initial phase of drilling, a device was used to stabilize the drilling position. During operation, the drill was kept at low speed so as not to damage the coating. The distance between the individual wells was at least 1 cm. After all holes had been drilled, the entire surface was dusted and washed with acetone-soaked sewage. Figure 3 shows a cross-section and a photo of the finished test disc. Then the surfaces of the steel discs were degreased with acetone for better adhesion. The discs were glued to the coating system with the Epidian 5 a)

b)

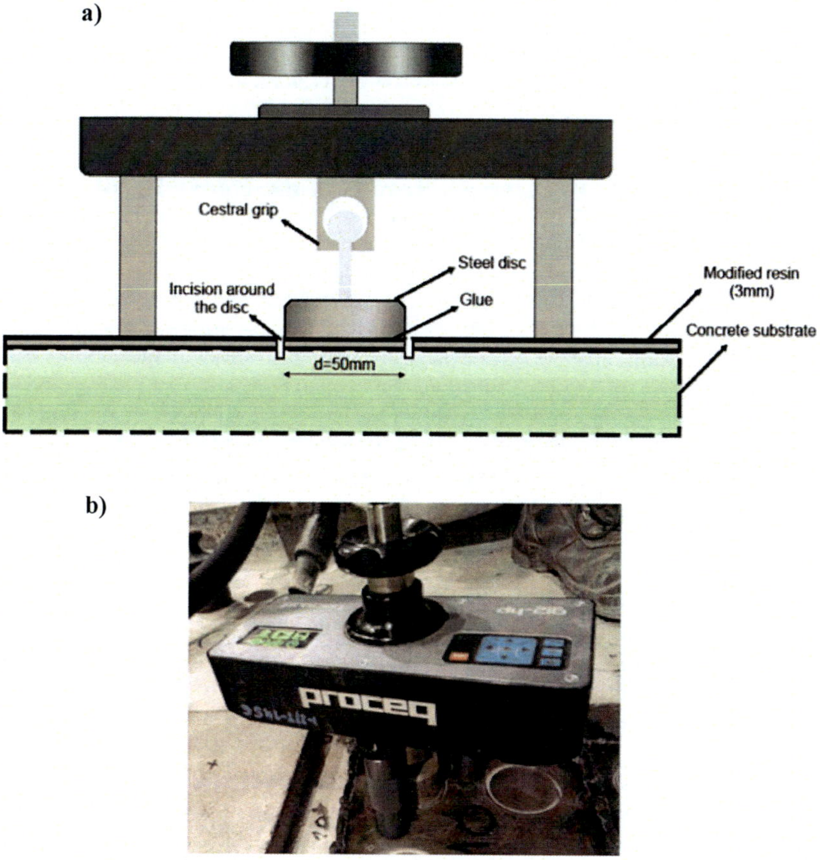

Fig. 3 Steel discs connecting the coating with an automatic device for testing the pull-off strength: **a** scheme of the test stand, **b** photo of the test stand with the device for testing the pull-off strength

epoxy resin with the Z1 hardener from Ciech Sarzyna (Nowa Sarzyna). The ratio was 100: 12.

The places where the measurements were taken had a flat surface, keeping the appropriate distance from the edge. The standard (ASTM D4541–17 2017) specifies that for the test to be reliable, more than three tests should be performed, spaced so that it is possible to disassemble the device. The glue was applied to both surfaces—the coating and the disc. After sticking, the excess glue that ran out of the anastomosis surface was removed. Constant pressure on the disc is ensured in the initial phase of the bonding of the glue. After the adhesive was fully cured (in this case after about three weeks), the procedure of testing the adhesion of the coating was started. The connector was screwed into the steel disc, which allowed the disc to be articulated with the jaw of the device. After the device was set up and leveled, a test was performed. It lasted up to 100 s. The load was increased at the rate of 0.05 MPa/s. The automatic device used in the study is the Proceq Dy-216. It recorded the result, along with a graph of the load increment over time.

3 The Pull-Off Strength Results of the Epoxy Resin Bonding Agent

In 11 tests out of 28, surface detachment of the coating substrate (adhesive type) was found: one each with the addition of 10% and in the reference test, two each with the addition of 30, 40 and 50%, and three fractures with the addition of 60%. In the majority of cases, the substrate was broken (cohesive type). As can be seen, the greater the amount of waste granite powder added, the more frequent the ahchezia detachment occurred. This may indicate that the resin, while still in a liquid state, was so dense that it was not able to reach every pore on the surface of the primer. As a result of what, the proper fusing of these two surfaces occurred, which can be seen from the types of detachment (as well as the pull-off strength).

Figure 4 shows a diagram of the dependence of the pull-off strength of the epoxy resin-based bonding agents on the concrete substrate depending on the amount of waste granite powder added. It shows that the best result in the test was 3.40 MPa, with the addition of 20% waste granite powder. This resulted in an increase in strength compared to the reference coating system (2.88 MPa) by 0.52 MPa, i.e., 18%. This shows that the strength can be improved with little effort. The second result in the test was obtained with the addition of 10% waste granite powder. The pull-off strength was 3.20 MPa—strength increase by 11% (0.32 MPa). When 30% of the powder was added, the test result was the lowest—2.49 MPa. A greater amount of waste granite powder resulted in a decrease in strength compared to the reference coating systems.

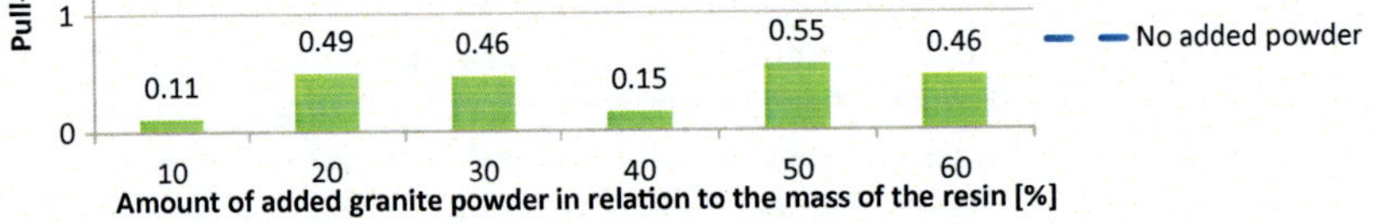

Fig. 4 The dependence of the pull-off strength of the epoxy resin-based bonding agents on the substrate depending on the amount of waste granite powder added

The summary of all results, together with the standard deviations of individual tests and the value declared by the manufacturer, is presented in the figure below.

Figure 5 shows photos of the shape of the detachment surface. In 11 trials, the subsurface surface was torn off. It is most visible in the reference coating system (Fig. 5a, marked with a red arrow). In essence, all peeled surfaces resulted in sharp edges and fine, loose fragments of the concrete substrate. No relationship was found between the shape of the detachment and the strength value.

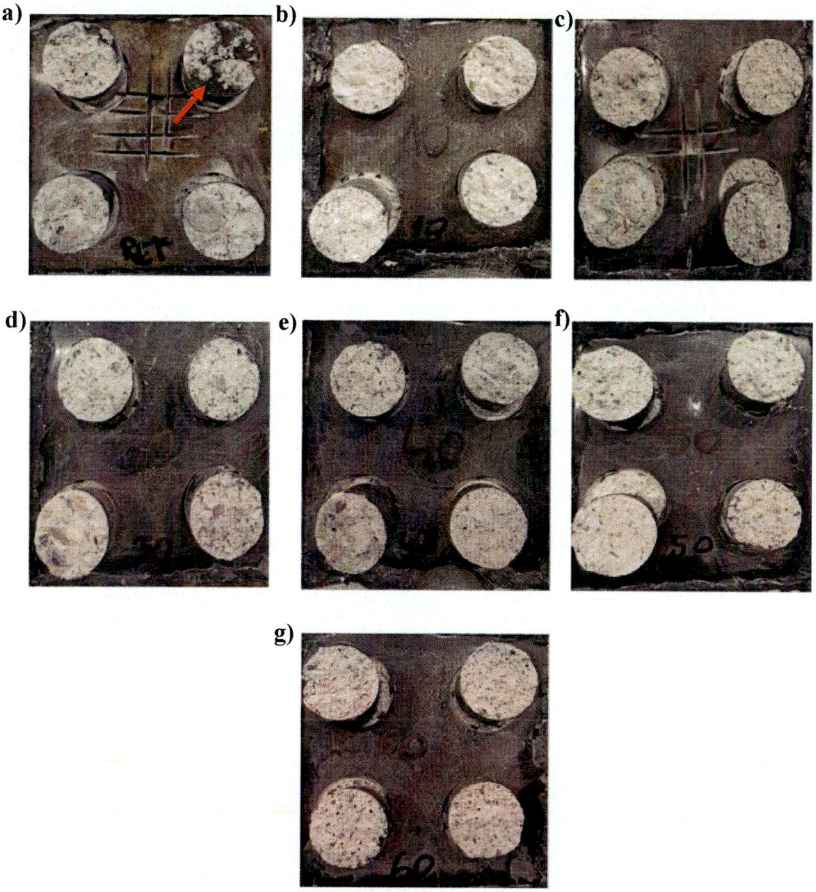

Fig. 5 Photos of epoxy resin-based bonding agent coating system after detachment: **a** without additive (one cohesive fracture), **b** 10 wt. % of the additive (one cohesive fracture), **c** 20 wt. % (all cohesive failure), **d** 30 wt. % (two cohesive fractures), **e** 40 wt. % (two cohesive fractures), **f** 50 wt. % (two cohesive fractures), **g** 60 wt. % (three cohesive fractures)

4 Cost-Economic Analysis

In view of the constantly rising prices, nowadays, savings are increasingly being sought in all industries, especially in the construction industry. The obvious approach is to use various types of cheaper alternatives, which at the same time will not reduce the properties of the finished product. The waste granite powder described in this article turns out to be a good example. As described above, an adequate amount may even increase the pull-off strength of the coating.

A cost-economic analysis was carried out to verify the main costs related to the coating. The focus was only on the direct purchase cost of epoxy resin and waste granite powder. The cost of making such a coating was not considered. The price of resin was assumed as 200 €/12.5 kg, and for waste granite powder 8.90 €/1000 kg. The breakdown of costs for the individual coating systems subjected to the strength test is presented below in Table 2. To compare the costs and the mechanical properties of the coating, their pull-off strength is also given.

The table clearly shows that the price per square meter decreases with the increasing amount of waste granite powder. In the best case, due to the pull-off strength of the coating (20% of the powder), it gives a final cost saving of 1.39 €/m^2, which means a cost reduction of 20%. This simple calculation clearly shows that it is easy to reduce the cost of making the coating and increase its pull-off strength. An additional advantage may be the color change to dark gray, which is a kind of advantage of the surface arrangement, no need to use dyes.

Table 2 A summary of the costs of making the tested coating depending on the amount of waste granite powder added

Amount of added granite powder in relation to the mass of the resin (%)	Resin content (%)	Epoxy resin cost (€/m^2)	Granite powder cost (€/m^2)	Total cost of an epoxy resin modified with waste granite powder (€/m^2)	Pull-off strength (MPa)
0	100	6.7	0.00	6.70	2.88
10	90	6	0.01	6.01	3.20
20	80	5.3	0.01	5.31	3.40
30	70	4.6	0.02	4.62	2.72
40	60	4	0.02	4.02	2.71
50	50	3.3	0.03	3.33	2.61
60	40	2.7	0.03	2.73	2.66

4.1 Mechanical and Cost Analysis of the Tested Coating

To compare the obtained results of the pull-off strength, the Mechanical Performance Ratio (MPR) was used

$$MPR = \frac{F_{mizes}}{F_{Ref.}} \times 100\%[\%] \quad (1)$$

To compare the Mechanical Performance Ratio (MPR), ΔMPR was calculated as follows:

$$\Delta MPR = \frac{MPR_{series}}{MPR_{Ref}}[-] \quad (2)$$

Figure 6 shows the value of the Mechanical Performance Ratio (MPR) in relation to the tested coating. The value of the MPR ratio mainly increases with the value of the addition of 20%. The value of the MPR index rises mainly with the value of the addition of 20%. With more granite powder, the value drops to about 90%

The cost of epoxy resin-based bonding agents depends mainly on the cost of components and location. To determine the value of individual material costs in relation to their importance for the composition of the epoxy coating, the Effective Cost Ratio (ECR) was used:

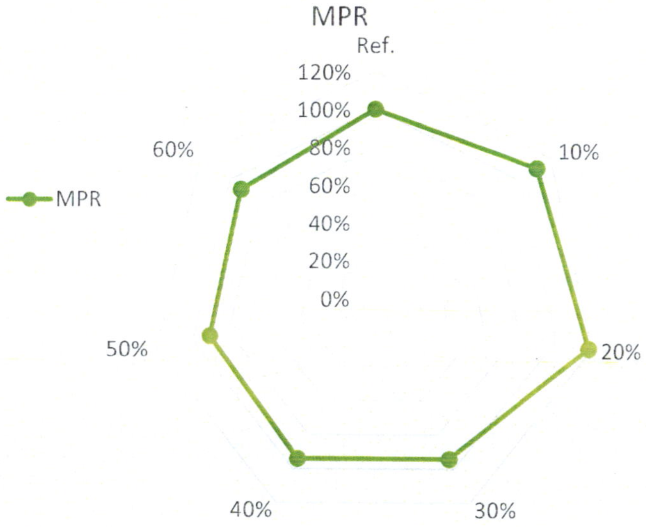

Fig. 6 Mechanical efficiency factor (MPR) in relation to the tested coating

$$ECR. = \frac{C_{mixes}}{C_{Ref}} \times 100\%[\%] \tag{3}$$

$$\Delta ECR. = \frac{ECR_{mixes}}{ECR_{Ref}}[-] \tag{4}$$

Figure 7 shows the value of the Effective Cost Ratio (ECR) for the coating tested. The ECR value decreases as the amount of waste granite powder in the coating increases. With a larger amount it is much lower than in the case of the reference coating systems, there is a linear relationship.

The results of both (MPR and ECR) factors are summarized in Table 3.

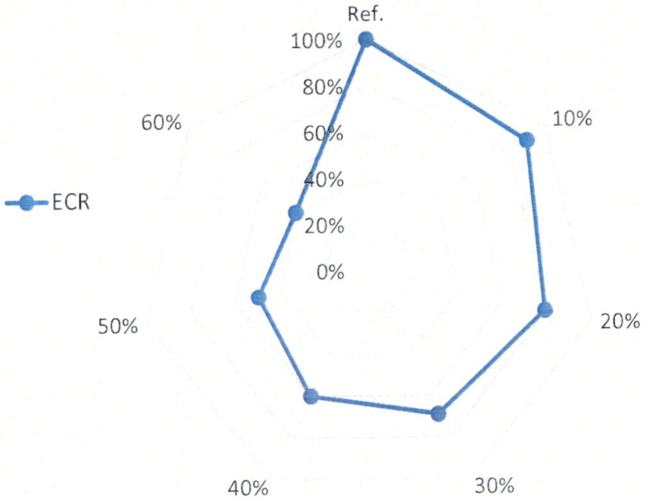

Fig. 7 The Effective Cost Ratio (ECR) of the tested coating

Table 3 The results of the analysis of the mechanical and economic properties for the tested coating

Amount of added granite powder in relation to the mass of the resin [%]	MPR [%]	Δ MPR	ECR [%]	Δ ECR
0	100	1.00	100	1.00
10	111	1.11	90	0.90
20	118	1.18	79	0.79
30	94	0.94	69	0.69
40	94	0.94	60	0.60
50	91	0.91	50	0.50
60	92	0.92	41	0.41

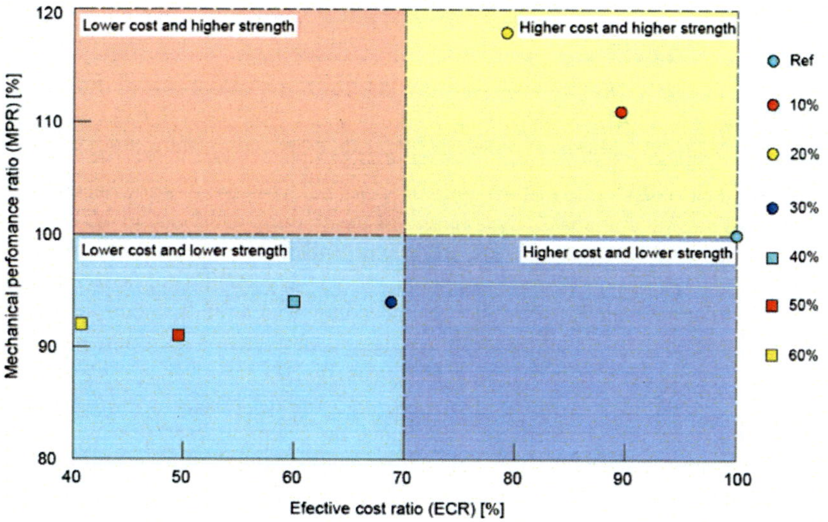

Fig. 8 Mechanical Performance Ratio (MPR) and Effective Cost Ratio (ECR) for the coating tested

Figure 8 presents the division of the tested coatings in relation to the Mechanical Performance Ratio (MPR) and the value of the Effective Cost Ratio (ECR) obtained in the study. The choice of the optimal solution should be a result of what goal you want to achieve: maximum cost reduction, maximum pull-off strength, or optimal value.

The above diagram shows that two additive values (and 10% and 20%) cause the highest strength gains, but the costs of their implementation are at the level of over 70% compared to the reference coating systems. This variant is recommended when a high-strength coating is required. In the "Lower Cost and Greater Endurance" area, no results qualified. The field with the most results ("Lower Cost and Less Strength") shows the results that do not make much sense to use. Yes, they can significantly reduce costs, but also reduce pull-off strength. The MPR value drops sharply after adding more than 20% of granite waste powder, later the values are almost linear. It is also significant that none of the results were also classified as "Higher cost and lower endurance", which is important information for future research on the subject.

5 Conclusions

Waste rock powders are the subject of various types of articles. However, the topic of connecting waste granite powder with epoxy resin is not anymore. This article examines the mechanical properties and performs an economic analysis of the modified coatings. The main aim of this study was to obtain the highest possible pull-off strength of the coating and to reduce costs as much as possible. Therefore, the test was carried out on a concrete substrate, where the influence of the addition of 10, 20, 30, 40%, 50, and 60% of waste granite powder to the epoxy resin primer was analyzed. The obtained results were compared with the reference coating system (without the addition of powder). The main conclusions of the study are:

- It has been found that the optimal content of waste granite powder is 10–20 wt.%. This value increases the strength of the coating, which increases the strength of the coating by up to 18% compared to the reference coating system. A larger amount of powder caused a reduction or a very small increase in strength,
- Economic analysis has shown that adding even a small amount of waste granite powder to the coating results in a significant reduction in coating costs. If 20% of the powder is added, it reduces the cost by 20%.
- The use of waste granite powder improves the properties of the coating. This gives a great opportunity to continue the research, inter alia, by modifying the composition of the substrate or the coating itself, in the way described in the introduction. This will allow to further increase the pull-off strength of the coating, reduce costs, and have a positive impact on the environment by eliminating the harmful waste granite powder.

Acknowledgements The authors thank Adrian Chajec for making the concrete substrate and Mateusz Moj for help in making the resin coatings.

Funding The authors received funding for the project supported by the National Centre for Research and Development, Poland, grant no. LIDER/35/0130/L-11/19/NCBR/2020 "The use of granite powder waste for the production of selected construction products."

References

Ahn, N.: Effects of diacrylate monomers on the bond strength of polymer concrete to wet substrates. J. Appl. Polym. Sci. **90**, 991–1000 (2003)

ASTM D4541–17: Standard test method for pull-off strength of coatings using portable adhesion testers. ASTM International: West Conshohocken, PA, USA, 06.02 1–6 (2017)

Atta, A.M., Al-Hodan, H.A., Hameed, R.S.A., Ezzat, A.O.: Preparation of green cardanol-based epoxy and hardener as primer coatings for petroleum and gas steel in marine environment. Prog. Org. Coat. **111**, 283–293 (2017)

Chajec, A.: Granite powder vs. fly ash for the sustainable production of air-cured cementitious mortars. Materials **14**(5), 1208 (2021)

Chi, J., Zhang, G., Xie, Q., Ma, C., Zhang, G.: High performance epoxy coating with cross-linkable solvent via diels-alder reaction for anti-corrosion of concrete. Prog. Org. Coat. **139**, 105473 (2020)

Chowaniec, A., Sadowski, Ł., Żak, A.: The chemical and microstructural analysis of the adhesive properties of epoxy resin coatings modified using waste glass powder. Appl. Surf. Sci. **504**, 144373 (2020)

Chowaniec, A., Ostrowski, K.: Epoxy resin coatings modified with waste glass powder for sustainable construction. Czasopismo Techniczne **8**, 99–109 (2018)

Do, J., Soh, Y.: Performance of polymer-modified self-leveling mortars with high polymer–cement ratio for floor finishing. Cem. Concr. Res. **33**, 1497–1505 (2003)

Francke, B., Piekarczuk, A.: Experimental investigation of adhesion failure between waterproof coatings and terrace tiles under usage loads. Buildings **10**(3), 59 (2020)

Kampa, Ł., Chowaniec, A., Królicka, A., Sadowski, Ł.: The effect of the addition of polypropylene fibers to primer on the pull-off strength of epoxy resin coatings. Materials, 4674 (2020)

Kim, S., Hong, H., Han, T.H., Kim, M.O.: Early-age tensile bond characteristics of epoxy coatings for underwater applications. Coatings **9**, 757 (2019)

Krzywiński, K., Sadowski, Ł: The effect of texturing of the surface of concrete substrate on the pull-off strength of epoxy resin coating. Coatings **9**, 143 (2019)

Li, L., Wang, Y., Tan, Y., Kwan, A.: Filler technology of adding granite dust to reduce cement content and increase the strength of mortar. Powder Technol. **342**, 388–396 (2019)

Morshed, S.A., Young, T.J., Chirdon, W.M., Zhang, Q., Tatar, J.: Durability of wet lay-up FRP bonded to concrete with nanomodified epoxy adhesives. J. Adhes. **96**, 1–26 (2020)

Othman, N.H., Yahya, W.Z.N., Che, I.M., et al.: Highly dispersed graphene oxide–zinc oxide nanohybrids in epoxy coating with improved water barrier properties and corrosion resistance. J. Coat. Technol. Res. **17**, 101–114 (2020)

Patil, D.M., Phalak, G.A., Mhaske, S.T.: Synthesis of bio-based epoxy resin from gallic acid with various epoxy equivalent weights and its effects on coating properties. J. Coat. Technol. Res. **14**, 355–365 (2019)

Pourhashema, S., Vaezia, M., Rashidib, A.: Investigating the effect of SiO_2-graphene oxide hybrid as inorganic nanofiller on corrosion protection properties of epoxy coatings. Surf. Coat. Technol. **311**, 282–294 (2017)

Rizwan, M., Khaleequzzaman, S., Khaleeq uz Zaman, U., Aqib Fida, S., Shahzad, A., Rehman, M., Oneeb Sulaiman, M.: Tensile strength improvement of concrete slabs using polyurea spray. Pract. Periodical Struct. Des. Construct. **26**(1) (2021)

Sadeka, D.M., El-Attarb, M.M., Alib, H.A.: Reusing of marble and granite powders in self-compacting concrete for sustainable development. J. Cleaner Product. (2016)

Sun, D., Hou, P., Li, B., et al.: Catechol-modified epoxy coatings with high adhesive strength on saturated concrete substrate. Iran. Polym. J. **30**, 401–410 (2021)

Mechanical and Adhesive Properties of Cement-Lime Plasters Modified with Waste Granite Powder

Marlena Rudner, Adrian Chajec, and Łukasz Sadowski

Abstract The granite industry produces a troublesome waste product—granite powder. It is created during the cutting or splitting of granite stones. Currently, its use in industry is not widespread. In addition, granite powder stored in landfills poses a threat to the environment and harms human health. It is also the intention of this study to analyse the mechanical, adhesion and absorption properties of cement-lime plaster modified with waste granite powder. It is also the intention to determine what percentage of granite powder addition will allow the plaster to achieve optimum. All results obtained will be compared to the values of a reference sample without waste granite powder and analysed. The first step was to take the best predicted percentage of the additive in the finished plaster mortar. Then the mortar properties such as capillary rise, density of dried hardened mortar, consistency of fresh mortar, volume density of fresh mortar, compressive strength and adhesion to substrate of hardened mortar and also absorption properties were investigated. From the results obtained, it was found that the addition of waste granite powder improves the mechanical properties and adhesion of cement-lime plaster. This reduces the volume of mortar, and thus the cost of plastering, while utilizing the waste material. The use of granite powder as an additive in cement-lime plaster is also a solution that is beneficial to the environment and human health.

Keywords Sustainable engineering · Cements composites · Plaster properties · Adhesion of plasters · Industrial waste

M. Rudner · A. Chajec (✉) · Ł. Sadowski
Wrocław University of Science and Technology, Wybrzeże Wyspiańskiego 27, 50-370 Wrocław, Poland
e-mail: adrian.chajec@pwr.edu.pl

Ł. Sadowski
e-mail: lukasz.sadowski@pwr.edu.pl

L. F. M. da Silva and R. D. Adams (eds.), *6th International Conference on Adhesive Bonding 2021*, Proceedings in Engineering Mechanics,
https://doi.org/10.1007/978-3-030-87668-5_3

1 Introduction

The topic of managing waste generated during the processing of construction materials is becoming more and more common (Damare et al. 2017; Jankowska-Renkas 2010). Public awareness of the consequences of environmental pollution is increasing. One of the by-product materials that pollutes the air, as well as the soil, and above all harms human health, is granite powder (GP). This powder is produced during processing (cutting and breaking) of granite stone. Its storage and disposal is not a small problem for the environment. Granite powder stored in huge heaps is often blown by the wind to the surrounding fields. It is its fine form that is the greatest danger, as under this form it is able to get into the smallest cracks and contaminate the environment.

There is also the problem of high CO_2 emission into the atmosphere (Benhealal et al. 2013) generated by the cement industry. Cement is the primary material used to make concrete and plaster mortar. Sources state that more than 60% of the total CO_2 emissions come from cement plants (Baker et al. 2009; Bosoaga et al. 2009). The main reason is the huge consumption of limestone raw materials and fossil fuels. So carbon dioxide is released during transportation, production and hydration of cement. One idea to reduce CO_2 emissions is to replace the volume fraction of dry cement with additives that allow maintaining or even improving the properties of the resulting mixture.

A study on the use of waste granite powder as a filler in cement can be found in the literature (Bosoaga et al. 2009). The author observed a significant increase in the compressive strength of concrete in the initial and final stages (Gupta and Vyas 2018). Such a result gives the prospect of using waste granite powder in cement industry. The use of this additive on a larger scale will lead to the utilization of the waste product, while reducing the CO_2 emissions of the cement industry. However, it is worth noting that granite powder for mortar modification purposes should be properly prepared by cleaning and grinding. The effects of using waste granite powder have not yet been studied for plaster mortars.

Exterior plaster, are primarily to protect the supporting structure from adverse weather conditions. Their task is also to present the building. Therefore, durability is of utmost importance for the building structure (Chew and Tan 2003; Gaspar and Brito 2008). Resistance to water penetration into plaster is a desirable phenomenon. Reducing water transport can reduce the risk of freezing/thawing of water in winter, which causes material deterioration (Kočí et al. 2013; Maravelaki-Kalaitzaki 2007). In addition, during acid rain, unwanted chemicals are able to penetrate the plaster structure, thereby exerting harmful effects on it (Pavlíková et al. 2011). It is important to note the problems encountered with plaster. Cracking, falling off the substrate or spalling (Fig. 1) are just the tip of the iceberg.

The causes of shrinkage cracks, i.e. cracks have already been known, however, it is difficult to find out what to do to prevent them. Researchers from the University of Technology and Humanities in Bielsko-Biała investigated the influence of polypropylene and polyacrylic fibres admixtures in renovation mortars on the formation of these

Fig. 1 Photographs of cracks and spalling on exterior plaster

cracks. In their work (Prasetya 2019) they treated these fibres as reinforcement. They obtained positive results for mortars with the addition of polypropylene fibre (with a circular cross section and also with a length of 12 mm). In the literature (Bachraczek et al. 2014) we can also find studies conducted on gypsum plasters with the addition of rubber particles from recycled tires. The aim of this study was to reduce the probability of plaster cracking in the first hours after curing. Ultimately, it became impossible to verify the improvement in crack resistance because the tests showed no cracking in any test. However, an increase in modulus indicates a decrease in hardness (with the addition of rubber particles), which may result in poorer crack resistance behaviour. Laboratory studies on the shrinkage of gypsum materials were also conducted by Taha Ashour and Wei Wu. In their study (Serna et al. 2012), they used three types of fibres: wheat straw, barley straw, and wood chips. The result of their study was the conclusion that the formation of shrinkage cracks decreases as the fibre content increases, while it increases when the soil content in the mortar increases.

The absorbency of mortar and its adhesion to the substrate are important parameters of plaster that can cause a lot of trouble. If the adhesion of plaster is too low, it will fall off the wall. That is why it is important to test the parameters of a given material as widely as possible. Remember that adhesion properties depend not only on the type of mortar used, but also on the base material. Both compressive and tensile strength is also an important factor. Today's construction industry strives for versatility of products. The resulting materials are expected to exhibit the best possible properties in all key areas.

Analysing the work of researchers, one notices attempts to obtain such parameters by creating composite materials. In publication (Ashour and Wu 2010), cellulose fibres were added to cement plasters and their physical and mechanical properties were studied. The study showed that the presence of fibres resulted in higher water absorption results in the samples. The bulk density of the material decreased and a decrease in the compressive strength of the tested plasters was also noted. The effect of lime and cellulose ether on the properties of the plaster mortar was also investigated (Hospodarova et al. 2017; Stevulova et al. 2021). These additives caused a decrease in compressive strength and adhesion of the mortar. However, the workability of the mixture was improved.

As can be seen, the modification of plasters with recycled materials is becoming more and more common and leads to a positive trend in the development of composite materials (Spychał 2020; Spychał 2015; Fořt et al. 2018). The number of studies conducted on cementitious mixtures with granite powder has been increasing for years. Vijayalakshmi and Sekar (2013) evaluated the use of granite powder as a fine aggregate variable in concrete production. The results showed that an addition of up to 15% was beneficial and had no adverse effect on the strength and durability of the mix. The work (Vijayalakshmi et al. 2018) also replaced cement or sand with granite waste. The results obtained are characterized by a significant increase in splitting tensile strength and flexural strength, while the value of bond strength decreased slightly compared to the reference test. In addition, it has been noted, studying the use of granite powder in self-compacting concrete (Sadek et al. 2016), a reduction in the energy used in operating the vibrators, thus reducing construction time. The influence of mortar additives themselves, due to their type has also been described in the literature (Kittll and Castro 1975), thus showing the important role they play in the modern development of the materials industry.

Considering the above examples, it was considered that the main objective of this paper is to find the safe application of granite waste powder in cement-lime plaster. The mechanical, adhesive and absorption properties of cement-lime plaster modified with the above waste product were investigated. The work began by determining the basic parameters of the plaster mix, such as consistency and volume density of fresh mortar. The mechanical properties such as compressive strength of the mortar and flexural tensile strength were then investigated. The peel strength of the specimens was also analysed through adhesion testing. The last part of the study is the water absorption properties. The study also aims to determine the optimum percentage addition of granite powder to the plaster mortar. The results obtained were compared with the values obtained on the sample without the addition of granite powder and analysed. The research gap related to the combination of waste granite powder with cement-lime plaster mortar is hereby filled.

2 Materials and Methods

2.1 Blend Design of Composites and Sample Preparation

The research program included evaluation of selected mechanical and absorption properties of cement-lime plaster using granite powder addition. A ready-prepared cement-lime plaster mortar for manual application was used. It consists of hydrated lime, cement, sand and additives. The grain size of the plaster mix is 1 mm. Bulk density of the dry product is about 1550 kg/m^3. The grade of ready dry plaster mortar is CS II. The declared compressive strength of the plaster after 28 days is over 2.5 N/mm^2 (manufacturer information). The bulk density of granite powder used was 1194 kg/m^3. Formulations were prepared from the presented ingredients, taking into account the best possible assumed percentage of granite powder addition. Its amount was 5, 7.5 and 10% in relation to the weight of ready dry plaster mortar (Table 1).

Two types of specimens were made: specimens in the form of a layer of plaster placed on different substrates and rectangular specimens with dimensions of 40 mm × 40 mm × 160 mm. The mechanical properties of the mortar, such as flexural and compressive strength, and adhesion properties were tested on the cuboid specimens. The substrates for testing the adhesion of mortar to masonry elements were Silka brick of class 15, solid brick of class 25 and TERMOton P + W clay block of class 15.

Mortar for the samples was mixed in a plastic container with a hand mixer for 3–5 min according to the technical card of the product used. Then, preparing rectangular specimens, the mortar was poured into moulds previously greased with release agent. When preparing masonry units, the materials were not primed. This is because there are indications that the application of a primer has a negative effect on the peel strength of the plaster. However, the substrate was thoroughly wetted and then plaster 10 ± 5 mm thick was applied to the largest flat surfaces of the elements (Fig. 2).

Table 1 Mix design proportions of the mortars

Series of mixes (–)	Water (kg/m)	Ready mix plaster (kg/m^3)	Granit powder (kg/m^3)
Ref	310	1550	0
5%	325.50	1550	77.50
7.5%	333.25	1550	116.25
10%	341.00	1550	155.00

Fig. 2 Prepared cement-lime plaster applied on: **a** TERMOton, **b** ceramic solid brick, **c** Silka brick

2.2 Determination of Consistency of Fresh Mortars

The consistency of fresh cement-lime mortar with granite powder was determined using the spreading table method according to the requirements of the PN-EN 1015-3 procedure. A mould is placed in the centre of the disc of the spreading table, then the mortar is introduced into it in two layers. Each layer should be compacted with at least 10 blows of the rammer. Then the excess mortar is collected and the free surface of the disc is wiped dry. After about 15 s, you need to remove the mold and shake the spreading table 15 times. The final step is to measure with a calliper the diameter of the spread mortar in two perpendicular directions. Finally, the average value of the two measurements must be calculated and the result obtained is the value of the spread of the test sample.

2.3 *Determination of Volumetric Density of Wet Mortar*

The volumetric density of the wet mortar was determined according to PN-EN 1015-6. The mortar, prepared according to the product technical sheet, was placed in a mould of a specified volume ($V = 740$ cm^3). The mixture was then compacted and the mass was determined using a laboratory balance. The read mass was divided by the volume to obtain the volumetric density value of the wet mortar. The measurements were repeated for each recipe.

2.4 *Determination of Basic Mechanical Properties of Hardened Mortars*

The basic mechanical properties of the hardened mortars were determined by performing destructive tests such as compressive strength and flexural tensile strength. The rectangular specimens with dimensions of 40 × 40 × 160 mm were stored in humid environment. After 28 days, the whole specimens were subjected to flexural tensile strength test. Then, the halved specimens were subjected to compressive strength test. The tests were conducted in the laboratory of Wrocław University of Science and Technology, on a testing machine in accordance with the requirements of PN-EN 196-1 and PN-EN 1015-11.

2.5 *Determination of Adhesion of Hardened Mortars to the Substrate*

Tests to determine the adhesion of hardened mortars were performed on samples from masonry units (Fig. 3). Plaster mortar was applied to the previously prepared, moistened substrate according to the established recipes. The samples were kept in a humid environment. Next, holes were drilled in the plaster to a depth that allowed drilling into the substrate material. In the next step, the substrate was thoroughly vacuumed and washed with acetone. The discs were glued to the prepared surface using a two-component epoxy adhesive with a strength greater than the adhesive strength of the mortar. The discs were peeled off after the adhesive had set completely. The measurement results were read directly from the pull-off tester. The tests were performed according to the requirements specified in PN-EN 1015-12:2016-08E.

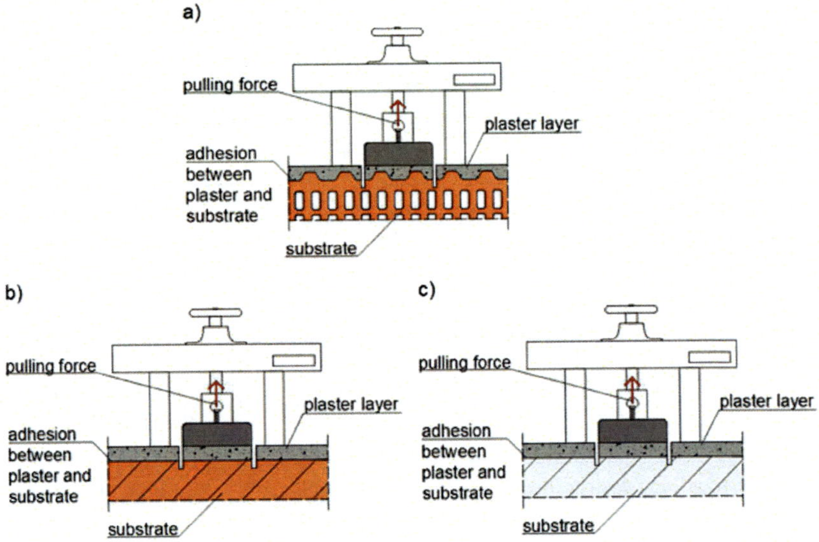

Fig. 3 Scheme of testing adhesion of hardened mortars to the substrate made of: **a** TERMOton, **b** ceramic solid brick, **c** Silka brick

2.6 Water Absorption Test

The water absorption properties were tested according to EN 1015:2002. Three rectangular specimens with dimensions of 40 mm × 40 mm × 160 mm were prepared. After 28 days of storage under humid conditions, the four long faces were sealed with liquid film and then each specimen was broken into two parts. The samples were then dried to constant weight at 60 ± 5 °C. Drying was completed when in two consecutive weighings (24 h interval) the weight loss was not greater than 0.2% of the total weight. After such preparation, the samples were placed in a cuvette, with the breakthrough surfaces of the cuboid towards the bottom (Fig. 4). A metal mesh was placed at the bottom of the cuvette to keep the rough surface of the samples entirely in water and to prevent air bubbles from being trapped underneath. The samples were submerged to a depth of 5–10 mm throughout the test. To maintain a constant water level and thus prevent evaporation, the cuvette was covered during the test.

Before placing the broken samples into the water, each sample was weighed, thus determining the initial sample mass. Once the samples were placed in the cuvette, a stopwatch was turned on and after a set time, each sample was weighed and then placed back in the cuvette. The times were: 5, 10, 15, 30, 45, 60, 90, 120, 180 and 240 min. Weighing was done after quickly wiping the wet surface with a damp cloth. The test was terminated when a clear wetting was observed on the free surface of the sample and it was ensured that the sample was completely saturated with water.

Fig. 4 Stand for testing absorption properties

Measurement of the weight of the sample after 10 and 90 min of soaking was used to determine the absorption coefficient. This coefficient was calculated from the formula:

$$C = 0.1 * (M2 - M1)\left[\frac{\text{kg}}{m^2 * \text{min}^{0.5}}\right] \tag{1}$$

M1 is the weight value of the sample weighed after 10 min of soaking in water. Whereas M2 is the mass of the same sample weighed after 90 min. These masses were expressed in grams. Before weighing the samples, wipe the bottom of the samples with a slightly damp towel.

The coefficients for each sample were calculated and then the average value of the water absorption coefficient C was determined. In addition, the average absorption rate for each series of samples was also calculated.

3 Results

3.1 Determining the Properties of Fresh Mortars

It was observed that the addition of waste granite powder had an effect on the wet mortar properties, the results are summarized below (Table 2). As the amount of additive increased, the diameter of the mix, as tested using a spreading table, decreased. The result for 10% granite powder addition was 152 mm, thus causing a 19.79% decrease in diameter compared to the reference sample. Comparing the values obtained for the reference sample and 5% powder addition, the volume density of the fresh mix increased by less than 0.5%. However, when comparing with 10%

Table 2 Properties of fresh mortars

Series of mixes (–)	Ref	5%	7.5%	10%
Spreading diameter (mm)	189.5	185.0	178.0	152.0
Volumetric density (g/cm^3)	1.692	1.700	1.767	1.864

addition, the value experienced an increase of more than 9% with respect to the reference sample.

In conclusion, the increase in bulk density of the fresh mixture is accompanied by a decrease in the diameter of the mortar tested with the spreader.

The increase in the volumetric density of the mixture with the increase in the amount of granite powder added is probably caused by a better filling of the voids in the mixture. The addition of granite powder improved the packing density of the mortar tested. The filling property of granite powder has already been reported in the literature (Li et al. 2019), and the results obtained confirm the tests carried out.

3.2 Determination of Basic Mechanical Properties of Hardened Mortars Modified with Granite Powder

The basic mechanical properties of cured mortars are the values of compressive strength and flexural tensile strength. A series of destructive tests were conducted to determine these parameters. These tests were carried out on cuboidal standardized specimens, 28 days after their execution.

The flexural tensile strength tests resulted in three results for each of the tested series of samples (Fig. 5a). In the measurements, no difference in strength was observed between the reference sample and the 5% granite powder addition. A clear change was noted for the sample with 10% addition, for which the flexural tensile strength averaged 8.75 MPa. Comparing to the reference sample, whose strength is equal to 3.75 MPa, thus an increase in strength values as much as 133.33%, was achieved. The intermediate result in the study of this parameter was determined for

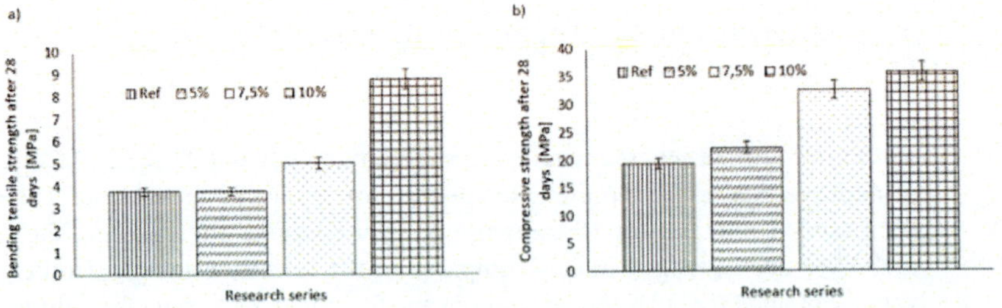

Fig. 5 Strength test: **a** bending tensile strength after 28 days (MPa), **b** compressive strength after 28 days (MPa)

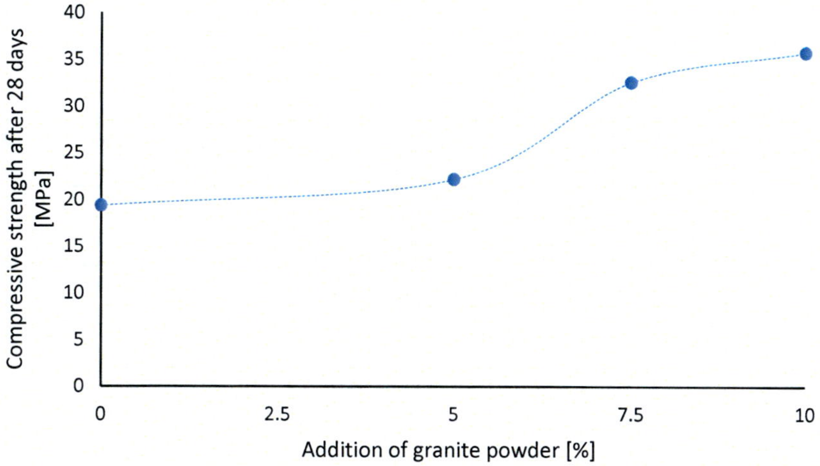

Fig. 6 Increase in compressive strength as a function of granite powder addition

the addition of 7.5% amounting to 5 MPa. Thus, it can be concluded that an increase in the amount of granite powder addition results in an increase in flexural tensile strength of hardened cement-lime plaster mortar.

Broken specimens were used to perform the compressive test, thus obtaining 6 measurements from each series, after the measurements were taken the results were averaged (Fig. 5b). An increase in strength was observed with an increase in the addition of granite powder. The graph (Fig. 6) shows that the increase in compressive strength was initially slow. Then comparing the same parameters obtained for 7.5 and 10% granite powder addition, a dynamic increase in the mechanical properties of the lime cement plaster is observed.

The results for the compressive strength parameter show a continuous upward trend. The initial increase in this parameter was small. However, the increase between the reference sample and the 10% addition of granite powder was as high as 84.33%. In the literature (Jacewicz-Rek et al. 2015), an increase in mechanical properties was also reported for waste ceramic plaster additive. The increase in compressive strength was finally 11% there. The results obtained also confirm the results obtained by the researchers (Medina et al. 2012), who achieved a 12% increase in compressive strength when replacing 25% of gravel with sanitary ceramic waste (Fig. 7).

The statement that the addition of granite powder improves the mechanical properties of the mix is also confirmed by the study (Jankowska-Renkas 2010) and other literature (Belebchouche et al. 2021). In the cited work, an increase in compressive strength was achieved after modification of concrete with chemical admixture and granite powder.

The improvement of mechanical properties of cement-lime plaster is related to packing-density phenomenon (Fig. 8). The fine grains of granite powder fill the free spaces in the mortar, thereby improving compressive and bending tensile strength.

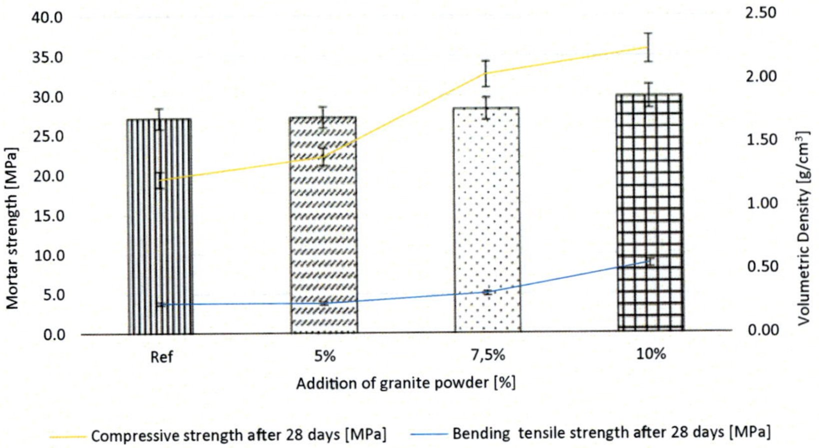

Fig. 7 The relationship between the volume density (g/cm^3) of a tested mortar and its strength (MPa)

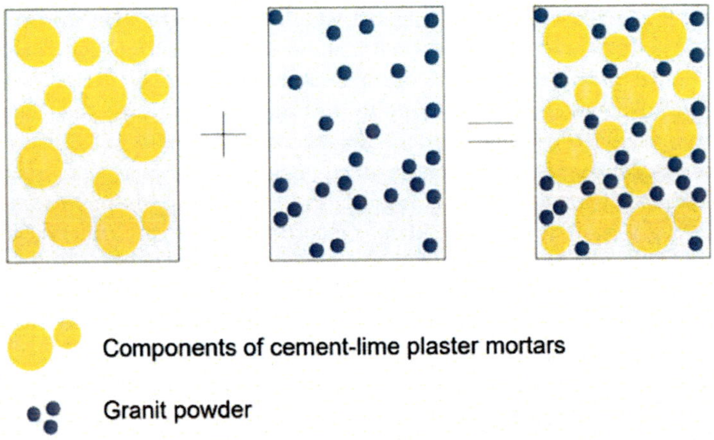

Fig. 8 Packing density

3.3 Determination of Adhesion of Hardened Mortars to the Substrate

To test the adhesion of hardened mortar to the substrate, cement-lime plaster with granite powder was applied to three types of substrates. The first was a TERMOton hollow block, the second a silicate block and the last an ordinary solid brick. These materials were chosen because of their popularity for use in general construction.

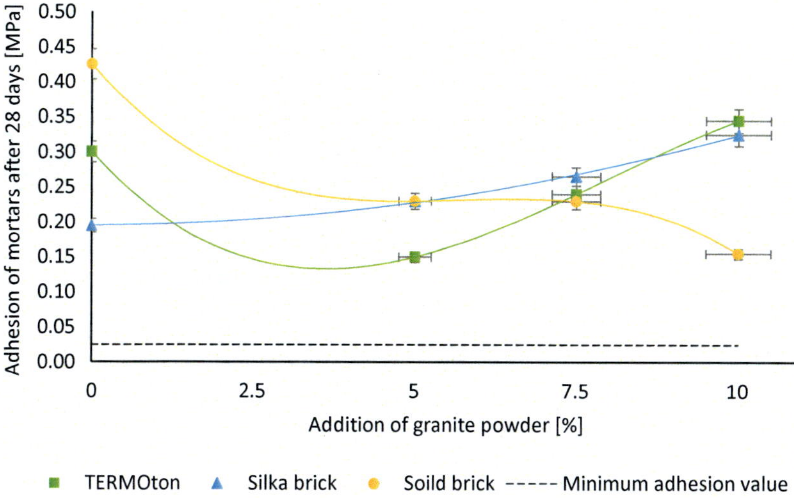

Fig. 9 Change in adhesion of hardened mortars to the substrate with respect to the type of substrate and the amount of granite powder added

The results obtained show significant differences in mortar adhesion due to the substrate used. For the reference sample, the adhesion values were 0.30 MPa for TERMOton, 0.195 MPa for silica and 0.425 MPa for solid brick. Then, for mortar with 5% addition of lime powder, a decrease in peel strength was recorded for TERMOton and solid brick substrates. The adhesion to Silka brick substrate still increased. The pattern of variation of adhesion of cement-lime mortar with granite powder addition to the substrate is shown in Fig. 9. It is worth mentioning that the minimum adhesion value for cement-lime plaster is 0.025 MPa.

It is worth mentioning that on the TERMOton substrate at 5% dry granite powder addition, the adhesion decreased drastically compared to the reference sample. However, it then increased as the amount of additive increased. Thus obtaining for 10% granite powder addition an increase in mortar adhesion of 15% with respect to the sample without additive.

When determining the mortar adhesion to the substrate by the pull-off method, a continuous decrease in this parameter occurred on the substrate made of solid ceramic brick. The decrease was also as high as 63.53% comparing the reference sample with the sample with 10% granite powder added. A decrease in adhesion can occur as a result of the difference in coefficients of thermal expansion of the mortar and substrate or the difference in modulus of elasticity E of the mortar and substrate.

The best results were obtained by testing the adhesion of the mortar to the substrate on Silka bricks. The adhesion value increased with an increase in the amount of granite powder added. At 10% addition, the tested value reached 166.67% of the reference sample value.

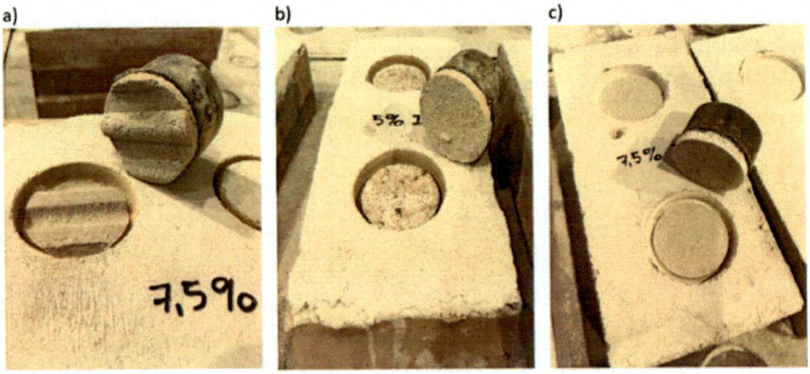

Fig. 10 Separation from the substrate made of: **a** TERMOton, **b** solid brick, **c** silka brick

The results obtained are in agreement with the results present in the literature (Niedostatkiewicz et al. 2019; Szymanowski 2019). The researchers obtained adhesion values of cement-lime mortar without additives in the range of 0.29–0.33 MPa depending on the substrate. Irrespective of the percentage of granite powder addition, mortar peeling was observed at the mortar-substrate interface (Fig. 10).

The increase in mortar adhesion to the substrate is probably related to the packing-density phenomenon. The granite powder tightly fills the spaces between the grains of the plaster mortar and at the same time increases the mortar's contact surface with the substrate (Fig. 11). This process thus improves the adhesion of the mortar.

Researchers (Bochen 2015) studying the effect of weather conditions on the properties of plaster mortars have also noted a correlation between high plaster strengths and better adhesion results. Their study also confirms the obtained dominant increase in adhesion on silicate blocks. This increase may be due to the binding of traditional plasters with the cement binder, or the lime contained in it.

3.4 Water Absorption Test

The water absorption test was carried out on the previously prepared cuboid beams. The obtained results of the samples after a certain time are shown in Fig. 12. The test was terminated after 2 h, because at that time water appeared on the upper surface of the specimen. Cement-lime mortars have an absorption coefficient ranging from 0.6 to 1.9 kg/m^2 $\times$ min 0.5 depending on the mortar variety.

Based on the results obtained, the value of the said absorption coefficient "c" was calculated using formulas. It is unequivocally stated that as the amount of granite powder addition increased, the value of this coefficient decreased. The results obtained for the different percentages of granite powder additions are summarized in Table 3. These results could be achieved through the previously mentioned role

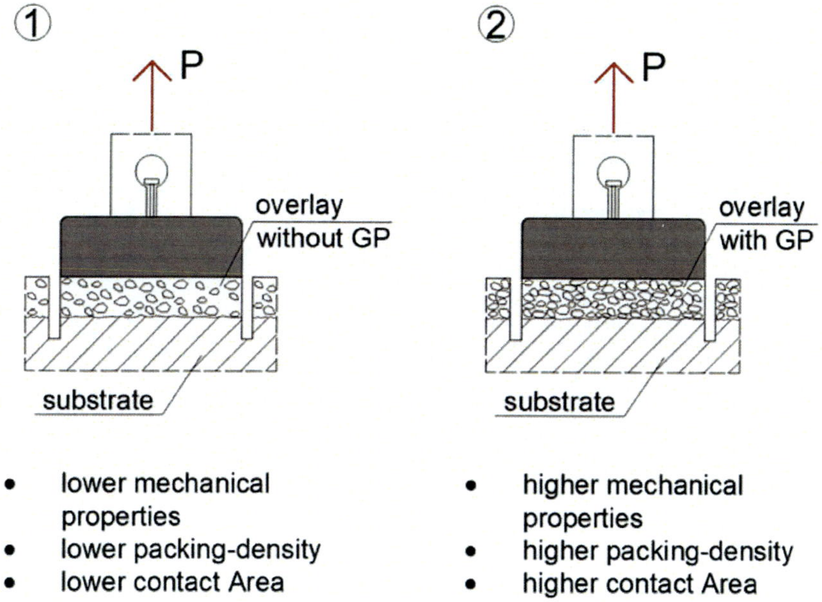

Fig. 11 Cross-section through the plaster mortar during pull-off test

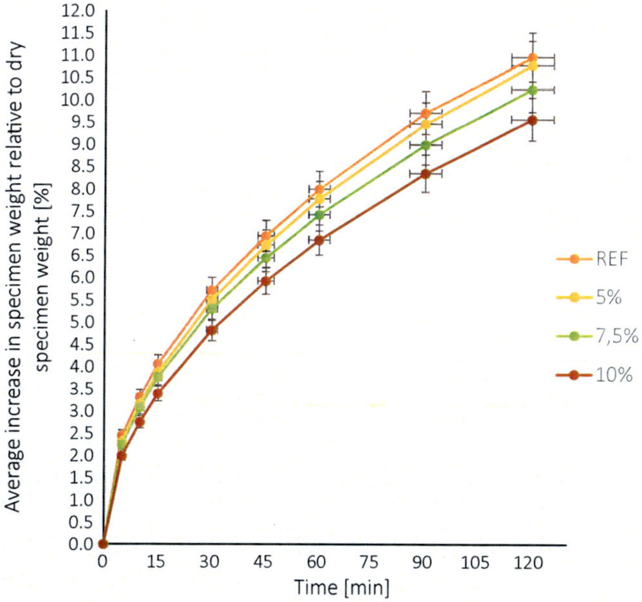

Fig. 12 Average percentage weight gain of the sample subjected to the water absorption test, after a specified time

Table 3 Properties of fresh mortars

Series of mixes (–)	Water absorption coefficient (–)
Ref	0.640
5%	0.631
7.5%	0.591
10%	0.560

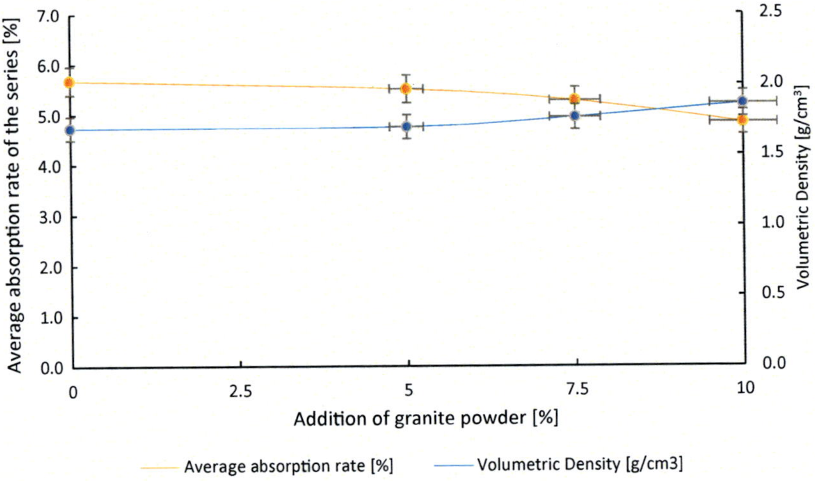

Fig. 13 Relation of bulk density to mean absorbability of a series

of granite powder as a pore filling material for the mortar. The more additive used in the sample, the heavier the sample was, which means a more compact mass.

Parameters such as the average absorbency for each batch and the volumetric density of the wet mortar are also worth comparing with each other (Fig. 13). As the addition of granite powder increases, the bulk density increases. The opposite is true for the average absorbency of the mortar. As a result of the spaces in the mortar being thoroughly filled, the water flow path becomes longer and thus the absorbency decreases.

3.5 *Comparison of Obrained Results and Literature Review*

Analysing the data collected in Table 4 (including our own results) it should be stated that the results for the granite powder additive, regardless of the mortar type, show similar trends in both adhesion, absorption and strength properties. Cellulose fibre additives to the plaster mortar caused a decrease in the compressive strength of the mortar as well as an increase in water absorption. These additives behaved in the

Table 4 Comparison of results obtained and literature review on mortar additives

Reference	The tested scale of composite	Type of additive	The properties of the mortar to be tested (development or regression)							
			Mechanical properties						Absorption properties	
			Compressive strength		Bending tensile strength		Peel-off strength		Absorbability	
			Reference series	Mortar additive	Reference series	Mortar additive	Reference series	Mortar additive	Reference series	Mortar additive
This research	plastering mortar	granite powder	100%	184,33%	100%	233,33%	100%	166,67% (silca brick)	100%	84,21%
Hospodarova et al. 2017	plastering mortar	cellulose fibers	100%	53,10%	-	-	-	-	100%	113,50%
Stevulova et al. 2021	plastering mortar	recycled cellulose fibers	100%	67,92%	100%	42,19%	-	-	100%	132,41%
Jackiewicz-Rek et al. 2015	mortar	cermic waste fillers	100%	111%	100%	112%	-	-	-	-
Gupta et al. 2018	mortar	granite powder	-	-	-	-	100%	123%	100%	94,80%
Jankowska-Renkas 2010	mortar	granite powder	100%	103,62%	-	-	-	-	-	-

\- lack of data

opposite way to the granite powder additive, which increased the mortar strength and contributed to a decrease in its absorbability. This increase in strength parameters can also be observed for the addition of ceramic waste. The data summarized in the table represent the most favourable results measured by each researcher.

4 Conclusions

The article referred to the description of tests conducted on cement-lime plasters with the addition of granite powder. It presents a description of the current state of knowledge about the possibilities of using waste granite powder in cement composites in a sustainable way. The results of testing the properties of fresh and hardened plaster mixtures are shown. After the analysis, it was found that the addition of granite powder can improve mechanical properties such as compressive strength and tensile strength of plasters. The volume density of fresh mortar also increases. The adhesion of cement-lime plaster mortar modified with granite powder to the substrate depends on the type of material used. This mortar performs best on Silka brick substrates. The plasters with the addition of granite powder showed a reduced water absorption coefficient, so there was an improvement in this parameter. Finally, it is concluded that the mechanical and absorption properties of the mix are improved due to the addition of granite powder.

From these results, it should be concluded that granite powder has great potential for use as a supplementary material in cement-lime plasters. Given the improvement in key mortar properties, further research should be directed towards determining the maximum percentage addition of granite powder.

Acknowledgements The authors would like to thank Mateusz Moj for his help in organizing the study and making the samples and Łukasz Kampa for helping to prepare the samples.

Funding Source The authors received funding from the project supported by the National Centre for Research and Development, Poland (Grant No. LIDER/35/0130/L-11/19/NCBR/2020). "The use of granite powder waste for the production of selected construction products."

References

Ashour, T., Wu, W.: An experimental study on shrinkage of earth plaster with natural fibres for straw bale buildings. Int. J. Sustain. Eng. **3**(4), 299–304 (2010)

Baker, D., Turner, S.A., Nepier-Moore, P.A., et al.: CO_2 capture in the cement industry. Energy Proc. **1**, 87–94 (2009)

Belebchouche, C., Moussaceb, K., Bensebti, S.E., et al.: Mechanical and microstructural properties of ordinary concrete with high additions of crushed glass. Materials **14**(8), 1872 (2021)

Benhealal, E., Zahedi, G., Bahadori, A.: Global strategies and potentials to curb CO_2 emissions in cement industry. J. Clean. **51**, 142–161 (2013)

Bochen, J.: Weathering effects on physical–chemical properties of external plaster mortars exposed to different environments. Constr. Build. Mater. **79**, 192–206 (2015)

Bosoaga, A., Masek, O., Oakey, J.E.: Technologie wychwytywania CO_2 dla przemysłu cementowego. Procedura Energetyczna **1**, 133–140 (2009)

Brachaczek, W., Górecka, I., Gondek, E., et al.: Wpływ domieszek włókien polipropylenowych i poliakrylowych na powstawanie rys skurczowych tynków renowacyjnych. Materiały Ceramiczne **66**(4), 451–462 (2014)

Chew, M.Y., Tan, P.P.: Facade staining arising from design features. Constr. Build. Mater. **17**(3), 181–187 (2003)

Damare, A., Gupta, P., Baweja, R.: Utilization of granite powder in cement mortar. Int. J. Res. Publ. Eng. Technol. **3**(6), 69–72 (2017)

Fořt, J., Keppert, M., Černý, R.: Application of cellulose fiber reinforcement in lime plasters. Cem. Wapno Beton **23** (2018)

Gaspar, P.L., de Brito, J.: Quantifying environmental effects on cement-rendered facades: a comparison between different degradation indicators. Build. Environ. **43**(11), 1818–1828 (2008)

Gupta, L.K., Vyas, A.K.: Impact on mechanical properties of cement sand mortar containing waste granite powder. Constr. Build. Mater. **191**, 155–164 (2018)

Hospodarova, V., Stevulova, N., Junak, J., et al.: Cement materials based on cellulosic fibers for plasters. Sel. Sci. Papers J. Civ. Eng. **12**(1), 37–46 (2017)

Jackiewicz-Rek, W., Załęgowski, K., Garbacz, A., et al.: Properties of cement mortars modified with ceramic waste fillers. Proc. Eng. **108**, 681–687 (2015)

Jankowska-Renkas, E.: Wpływ mączki granitowej jako wypełniacza w cemencie na właściwości betonu. Roczniki Inżynierii Budowlanej **10**, 19–23 (2010)

Kittl, P., Castro, J.H.: Mortar cohesión. The effect of additives. Materiales de Construcción **25**(160), 27–30 (1975)

Kočí, J., Maděra, J., Rovnaníková, P., et al.: Hygrothermal performance of innovative renovation renders used for different types of historical masonry. Heritage Masonry Mater. Struct. **121** (2013)

Li, L.G., Wang, Y.M., Tan, Y.P., et al.: Filler technology of adding granite dust to reduce cement content and increase strength of mortar. Powder Technol. **342**, 388–396 (2019)
Maravelaki-Kalaitzaki, P.: Hydraulic lime mortars with siloxane for waterproofing historic masonry. Cem. Concr. Res. **37**(2), 283–290 (2007)
Medina, C., Sánchez de Rojas, M.I., Frías, M.: Reuse of sanitary ceramic wastes as coarse aggregate in eco-efficient concretes. Cem. Concr. Compos. **34**, 48–54 (2012)
Niedostatkiewicz, M., Majewski, T., Burdziński, M.: Tynki w budownictwie ogólnym -wybrane badania oraz problemy eksploatacyjne 33–35
Pavlíková, M., Pavlík, Z., Keppert, M. et al.: Salt transport and storage parameters of renovation plasters and their possible effects on restored buildings' walls. Constr. Building Mater. **25**(3), 1205–1212 (2011)
PN-EN 1015-12:2016-08E: Methods of test for mortar for masonry—Part 12: Determination of adhesive strength of hardened rendering and plastering mortars on substrates
Prasetya, W.H.: Inovasi high early strength concrete dengan pemanfaatan limbah batu granit, cangkang kerang dan fly ash. Jurnal Proyek Teknik Sipil **2**(2), 31–41 (2019)
Sadek, D.M., El-Attar, M.M., Ali, H.A.: Reusing of marble and granite powders in self-compacting concrete for sustainable development. J. Clean. Prod. **121**, 19–32 (2016)
Serna, Á., del Rio, M., Palomo, J.G., et al.: Improvement of gypsum plaster strain capacity by the addition of rubber particles from recycled tyres. Constr. Build. Mater. **35**, 633–641 (2012)
Spychał, E.: The effect of lime and cellulose ether on selected properties of plastering mortar. Proc. Eng. **108**, 324–331 (2015)
Spychał, E.: The rheology of cement pastes with the addition of hydrated lime and cellulose ether in comparison with selected properties of plastering mortars. Cem. Wapno Beton (2020)
Stevulova, N., Vaclavik, V., Hospodarova, V., et al.: Recycled cellulose fiber reinforced plaster. Materials **14**(11), 2986 (2021)
Szymanowski, J.: Evaluation of the adhesion between overlays and substrates in concrete floors: Literature survey, recent non-destructive and semi-destructive testing methods, and research gaps. Buildings **9**(9), 203 (2019)
Vijayalakshmi, M., Sekar, A.S.S.: Strength and durability properties of concrete made with granite industry waste. Constr. Build. Mater. **46**, 1–7 (2013)

Characterization of Fast Adhesive Curing Reactions—A Novel Experimental Setup

Sarah Glaser, Elisabeth Stammen, and Klaus Dilger

Abstract The desire to accelerate adhesive application processes drives the development of rapid curing adhesives. When selecting a certain adhesive for an application, typically a variety of tests is performed in order to characterize its properties. Especially, the mechanical properties and their development under different heating conditions are of interest. Commonly used analytical methods only provide limit information for rapid curing adhesives. This motivates the development of a new experimental setup enabling rheological measurements at high heating rates. This contribution defines requirements for such a novel setup. Induction heating is chosen to facilitate rapid temperature increase. Further, different materials are evaluated for their heating performance and their use for measurement components. A real-time capable temperature control is enabled due to the selected temperature sensor types and positioning. This study develops the design of heating units which can be mounted on a modern rheometer. Additionally, the impact of rapid heating on the measurement accuracy is investigated. Two major factors are identified: the influence of rapid temperature increase on the thermal expansion of the measurement geometry and on the torsional compliance.

Keywords Rapid curing · Rheology · Adhesives · Dynamic heating · Experimental setup · Compliance · Thermal expansion

S. Glaser (✉) · E. Stammen · K. Dilger
Institute of Joining and Welding, Technische Universität Braunschweig, Langer Kamp 8, 38106 Braunschweig, Germany
e-mail: sarah.glaser@tu-braunschweig.de

E. Stammen
e-mail: e.stammen@tu-braunschweig.de

K. Dilger
e-mail: k.dilger@tu-braunschweig.de

L. F. M. da Silva and R. D. Adams (eds.), *6th International Conference on Adhesive Bonding 2021*, Proceedings in Engineering Mechanics,
https://doi.org/10.1007/978-3-030-87668-5_4

1 Introduction

Understanding the curing mechanisms of adhesives presents basic knowledge for planning adhesive application and bonding processes. Thermal analysis—e.g. differential scanning calorimetry provides general information on the temperature range of the curing reaction. Rheological measurements make viscosity and modulus changes visible. The measured properties allow the simulation of the curing reaction (Jennrich et al. 2020, 2021). The demand for short manufacturing cycle times drives the development of rapid curing adhesives. Furthermore, the processing parameters influence both: the process time and the bond properties. As rule of thumb—based on the van-'t-Hoff rule—a temperature increase of 10 K decreases the curing time by half. In order to either trigger the start of a curing reaction precisely or to decrease the curing time, certain adhesives are formulated to cure at elevated temperatures.

The characterization and testing of adhesives at a fully cured state is well known (Da Silva et al. 2011, 2012; Chaves et al. 2014). Alterations of behavior during application and manufacturing processes, such as changes in viscosity, modulus or pot life are of interest, when newly selecting an adhesive for a certain application. Additionally, the curing properties are of a key interest, since they affect the quality of the bond. Commonly, for investigation of the curing behavior the adhesive is cured varying the curing parameters. It is then analyzed at a fully cured stage. Zimmer et al. (2017) investigates the cross-linking of polyurethanes using different methods of spectroscopy in order to gain information on the chemical composition. Mechanical properties are only tested at the fully cured stage. Rheological characterizations of adhesives commonly focus on: curing at isothermal conditions (Lucio and de La Fuente 2014; Papadopoulos et al. 2008) or gelation and vitrification studies at also isothermal conditions (Rochery and Lam 2000; Muller et al. 1991). Sánchez Cebrián et al. (2013) and Frauenhofer et al. (2012) elaborate on the desire of understanding the curing process more in depth. Especially, when it comes to heat curing adhesives for quality control purposes more information is needed. Rheological measurements generally allow monitoring the curing reaction. However, when looking at the rheology of fast curing adhesives, it is not possible to wait for steady-state conditions at an elevated temperature. An undefined proportion of the curing reaction will already have finished. The rheological insitu-monitoring of rapid curing reactions proposes experimental challenges since it violates the widely emphasized isothermal or near isothermal, hence steady-state measurement conditions. This motivates the development of a novel experimental set-up while revisiting these challenges. First, the fundamentals of rheological measurements, induction heating and fast curing adhesives are revisited, followed by a definition of requirements and measurement conditions for incorporating an induction heating unit into a modern rheometer. Subsequently, the design and testing of the prototype will be discussed.

1.1 Rheology

Usually, a disposable plate-plate configuration is selected as a measurement setup when investigating curing reactions. In this case the viscoelastic properties—determined from oscillatory tests—are of great interest. Curing behavior can be derived from the evolution of the complex shear modulus G^* where the real part denotes the elastic shear modulus $G^{'}$ and the imaginary part denotes the viscous shear modulus $G^{''}$. During the test the adhesive sample is exposed to a harmonic variation of either the shear stress τ or the deformation γ. The respective other is measured. δ represents the phase shift between excitation and response. The following equations show the mathematical relations (DIN 53019-4):

$$\gamma(t) = \gamma_0 e^{i\omega t} \tag{1}$$

$$\tau(t) = \tau_0 e^{i(\omega t+\delta)} \tag{1}$$

$$G^*(t) = \frac{\tau(t)}{\gamma(t)} = \frac{\tau_0}{\gamma_0}e^{i\delta} = \frac{\tau_0}{\gamma_0}\cos(\delta) + i\frac{\tau_0}{\gamma_0}\sin(\delta) = G' + iG'' \tag{3}$$

For a plate-plate configuration the deformation and shear stress differ along the radius of the plate (DIN 53019-1, Mezger 2016). Hence, the mean is defined at 2/3 of R_{max}, leading to the following expressions:

$$\tau_m(t) = \frac{4M_0(t)}{3\pi R_{max}}e^{i(\omega t+\delta)}\gamma_m(t) = \frac{2R_{max}\varphi_0(t)}{3h}e^{i\omega t} \tag{4}$$

$$G^* = \frac{2h}{\pi R_{\max}^4}\frac{M_0(t)}{\varphi_0(t)}(\cos\delta + i\sin\delta) = C_g\frac{M_0(t)}{\varphi_0(t)}(\cos\delta + i\sin\delta) \tag{5}$$

where $M_0(t)$ and $\varphi_0(t)$ are the torque and angular displacement over time, respectively. Furthermore, h is the gap between the two measurement plates. It is defined with regard to the investigated adhesive, especially with regard to the size of additives. $\frac{2h}{\pi R_{max}^4}$ is represented by the geometry factor C_g.

When analyzing the curing behavior of adhesives the measuring temperature and the heating rate substantially affect the materials behavior and the measurement accuracy. DIN 53019-4, ASTM D4473-08 and DIN EN ISO 6721 all emphasize the importance of steady state temperature conditions to inhibit a temperature gradient within the sample.

Heating rates of 0.5–5 K/min are recommended. Most rheometer manufacturers make provisions for a constant compensation factor to offset thermal expansion of the measurement geometry and enhance accuracy when applying heating rates. However, rapid heating processes are not intended. Peltier units allow a temperature range of −40 to 200 °C capable of heating rates of up to 30 K/min (Netzsch 2021).

These values are most commonly used for temperature driven plate-plate rheological measurements.

1.2 Basics of Induction Heating

Compared to external heating sources the internal generation of heat through induction achieves higher efficiencies and heating rates (Lupi et al. 2015). Michael Faraday discovered in 1831 that an alternating magnetic field of a primary conductor induces a current in a second wire, where there is no contact between the two conductors (Peter 2014). This basic principle of induction is often compared with the operating principle of an electronic transformer (Kegel 2013). The physical laws behind the induction phenomenon are described by Maxwell's equations, Faraday's law and the law of Joule-Lenz (Lupi et al. 2015). The magnetic field generates eddy currents within an electrically conductive material causing the component to heat up. The heat transfer (Eq. 6) can be derived from the first law of thermodynamics (Baehr and Stephan 2004).

$$\rho c(\vartheta)\frac{\partial \vartheta}{\partial t} = div[\lambda(\vartheta) grad\vartheta] + p_v \tag{6}$$

$$p_v = \frac{1}{2\kappa}|J(f, B)|^2 \tag{7}$$

where ρ is the mass density, c the specific heat, ϑ the temperature field, λ the thermal conductivity and p_v the heat source distribution. The latter represents the energy generated in the material and is put into mathematical context through (Eq. 7). The current density J within the heated part depends on the excitation frequency f of the magnetic field as well as the magnitude of the magnetic flux density B and the electrical conductivity κ of the heated materials. Since the influence of the magnetic field decreases with increasing distance between the inductor coil and the heated part these components shall be positioned as closed to each other as possible. Further, the permeability of the selected materials indicates the inductive effects. Additional heat can be generated through hysteresis loss. This is the case for ferromagnetic steel (Lupi et al. 2015; Küpfmüller et al. 2013; Fisk 2014).

1.3 Adhesive Curing Reactions

Depending on the chemical base of an adhesive there is a wide variety of cross-linking reactions and times. When looking at polyurethane adhesives, two component adhesives commonly cure faster than those with only one component. This is based on the diffusion controlled nature of the reaction of single component polyurethane adhesives. Moisture from the air surrounding the bond diffuses into the adhesive, curing it from the surface to the center. In order to accelerate the curing reaction while maintaining the advantage of a long open time several one component systems are formulated to incorporate a latent accelerator which may be triggered e.g. through a certain ambient temperature. This type of chemical formulation enables to start the curing reaction when desired, even within the open time (Cognard 2006).

Typically, the curing parameters can be found in the material data sheets of the adhesive. Differential scanning calorimetry (DSC) produces additional data, such as the initial temperature and the temperature range of the curing reaction at various heating rates. Two one component polyurethane adhesives with heat curing features were selected for preliminary studies.

1.3.1 Sikaflex®-360 HC

Sikaflex®-360 HC (Sika Deutschland GmbH, Stuttgart, Germany) is a stable one component polyurethane adhesive that cures to an elastomer. According to the technical data sheet (Sika 2011), the adhesive should not be cured with air humidity alone as the final properties and curing time may vary. In addition to moisture controlled curing, this adhesive also cures through a heat controlled reaction. Typical curing conditions of this adhesive are approximately 10 min at 150 °C. Due to its chemical structure, the adhesive is capable of building up a polymeric network with highly elastic properties through a number of cross-linking reactions. These are described in detail in (Jennrich et al. 2021).

DSC measurements show the start of the curing reaction at 100 °C for the heating rate of 10 K/min. With increasing heating rate the start of the reaction shifts to higher temperatures. This shift results from the thermal inertia. Due to its low thermal conductivity the sample's reaction lags behind the temperature.

1.3.2 Teroson 1510 PU

Teroson 1510 PU (Henkel AG & Co. KGaA) is a one component polyurethane adhesive. It is stable at room temperature and shall be cured at elevated temperatures. Compared to Sikaflex®-360 HC, which allows heat curing, this adhesive is specially developed for rapid heat curing processes. The data sheet explicitly mentions induction heating as one possible curing method (Henkel 2014).

DSC measurements show the start of the curing reaction at 90 °C for a heating rate of 10 K/min. Due to the sample's low thermal conductivity the curing reaction shifts to higher temperature levels with increasing heating rates, similarly as for the previous adhesive.

Additionally, the total reaction enthalpy increases with increasing heating rates which may suggest a change within the cross-linking structure to the product.

2 Development of a Novel Experimental Setup

Since an induction heating system for plate-plate rheology is not commercially available, a novel setup was developed. A patent from 1989 (Deer 1989) describes a possible setup without discussing the impact of the design on the measurement accuracy. Due to the physical structure and its geometrical constraints a different approach was needed to incorporate the induction heating into a modern readily available rheometer (Anton Paar MCR702 device).

It enables several test setup scenarios: induction heating of the upper or lower plate, only, heating both plates or combining an induction heating unit with a peltier unit. In compliance with (VDI 2019a, b) the development process was divided into three main stages—profile phase, concept phase and prototype phase.

2.1 Profile Phase

This phase defines the requirements, aims and properties of the novel experimental setup. The context of the development explained in the introduction section defines the basis of this phase. With regard to adhesive dependent requirements the adhesives presented previously shall provide input information. Furthermore, solution principles for certain design functions are elaborated and pretested.

2.1.1 Heating Rate and Temperature Range

DSC measurements and adhesive data sheets indicate the mayor curing reaction takes place within a temperature range of 100–160 °C. Thermogravimetric measurements display the risk of degradation of the adhesive above 200 °C. Thus, an operating range from room temperature to a maximum of 200 °C is defined for the induction heating unit. Induction heating may realize very high heating rates. However, as observed also from the DSC measurements (see Sect. Adhesive Curing Reactions) the thermal inertia of the adhesive must be taken into account. Hence, it is defined that a set temperature within the temperature range of the curing reaction shall be reached at a rate of 300 K/min or more.

2.1.2 Geometry and Materials

Geometrical constraints are defined through the connectivity options to the MCR702 device. Additionally, the plate-plate measurement configuration defines the geometry of the components in contact with the adhesive sample. Further, the plate-plate components need to be interchangeable. The increase of the torque during hardening of the adhesive limits the diameter of the sample. Measurement experience from (Jennrich et al. 2021) show, that plates with a diameter within the range of 8 or 15 mm are functional. This also satisfies the aspect ratio of plate diameter to measurement gap, which shall range from 10 to 50 according to DIN EN ISO 6721. Pretests were performed to assess the heating behavior of different non-corrosive materials with both inner field and traverse field inductors. Non-ferromagnetic; ferromagnetic steels and aluminum were compared. Since the inductive heat within the adhesive itself is negligible, the heat flux from the plates to the sample will trigger the curing reaction. The selected materials shall enable a homogeneous heat flux within the area of contact. Preliminary studies also show that the operating distance *d* between the heated component and the inductor shall be minimized. This observation is explained through the mathematical relation $B = \mu 0 \mu r H(I, d)$ of the magnetic flux density *B* within the heated part and the external magnetic field H. μ_0 represents the permeability constant and μ_r the relative permeability which depends on the given material (Fisk 2014). The latter indicates, and pretests confirm, that ferromagnetic materials with $\mu_r \gg 1$ are suitable for this application.

2.1.3 Temperature Measurement and Control

In order to relate the rheological behavior to the sample temperature it is desired to measure the temperature as close to the sample as possible. For a rapid temperature control process the temperature sensor shall be positioned as close to the heat source as possible. These two requirements need to be balanced (see Sect. Concept Phase). The control process is programmed with LabView. It incorporates a real-time capable proportional–integral–derivative (pid) controller. ASTM D4473-08 demands temperature stability within 1 K. It is aimed to realize this standard. However, an initial overshoot of 5 K shall be permitted, when first reaching the set temperature. In terms of temperature sensors thermocouples type K, which are not impacted by the frequency range of the magnetic field, or pyrometers shall be used. The measurement frequency of the sensors must support real-time control.

2.1.4 Measurement Conditions

Subsequently, to calculate the materials properties from the raw data, the machine errors need to be taken into account. Several researchers (Liu et al. 2011; Gottlieb and Macosko 1982; Läuger and Stettin 2016) have elaborated on the influence of instrument inertia and torsional instrument compliance. Laukkanen (2017) emphasizes the significance for the measurement results with an expected viscosity change by several orders of magnitude. The sample modulus G^*_s, G'_s and G''_s may be calculated from the measured modulus data (Eq. 5)—indicated now by the subscript *m*, where *J* represents the torsional compliance.

$$G_s^* = \frac{G_m^*}{\left(1 - \frac{J}{C_g} G_m^*\right)} \tag{8}$$

$$G_s^{'} = \frac{G_m^{'}\left(1 - \frac{J}{C_g} G_m^{'}\right) - \frac{J}{C_g} G_m^{''2}}{(1 - \frac{J}{C_g} G_m^{'})^2 + (\frac{J}{C_g} G_m^{''})^2} \tag{9}$$

$$G_s^{''} = \frac{G_m^{''}}{(1 - \frac{J}{C_g} G_m^{'})^2 + (\frac{J}{C_g} G_m^{''})^2} \tag{10}$$

Generally, J is simply the ratio $\frac{\theta}{M}$ of the angular deflection θ at a certain torque M. It can be determined experimentally through fixing one end of the measurement geometry and applying either a torque or a deflection at the other end (Schröter et al. 2006). With regard to the rheological results it needs to be taken into account that J varies depending on the measurement method and setup. In case of the endeavored setup, J is expected to change throughout the experiment due to the rapid increase of temperature. For oscillatory measurements, the instrument inertia may affect the measurement results.

According to Ewoldt et al. (2015) it specially proposes challenges for high-frequency data and for short-time data. For all materials the linear-viscoelastic region is determined prior to curing property measurements. These pretests have identified 1 Hz as a sufficient measurement frequency for the selected group of adhesives, hence leaving out high-frequency data. Additionally, oscillation shall commence shortly prior to the heating so that transient effects can settle and may be distinguished from effects caused through temperature change. Läuger and Stettin (2016) emphasizes the importance of instrument inertia, when measuring samples of low viscosity. This is also not the case for the selected group of adhesives which have a paste-like initial viscosity. Thus, it shall be relied on the use of a modern rheometer already including inertia correction options. Nevertheless, this approach can be verified by monitoring the signal of the encoder, as the authors mention.

ASTM D4473-08 indicates starting values for the measurement gap. For further considerations within the scope of this paper the gap is defined as 1 mm. Modern rheometers make provisions for a gap change due to thermal expansion of the measurement geometry. For steady state tests the deviation is calculated with the help of a constant pre-set expansion factor. Regarding the developed setup, the influence of thermal expansion on the gap size shall be evaluated, especially amidst the dynamic character of rapid curing reactions.

2.2 *Concept Phase*

During the concept phase solution concepts are generated based of the previous phase and advanced in detail as well as integrated into an experimental setup as a whole.

Figure 1a, b show the design of the induction heating units for the upper and lower plate, respectively.

A series of pretests lead to the presented design. Due to the geometry of the measurement shaft holding the upper plate, an inner field inductor was chosen for the upper unit, whereas a traverse field inductor was chosen for the lower unit. After identifying proficient induction parameters in terms of excitation frequency and pulse width modulation, respectively excitation current, the heating process further developed. Initial tests showed that the heating of the sample will be enabled through heat transfer from the metal plates of the rheometer to the adhesive. The

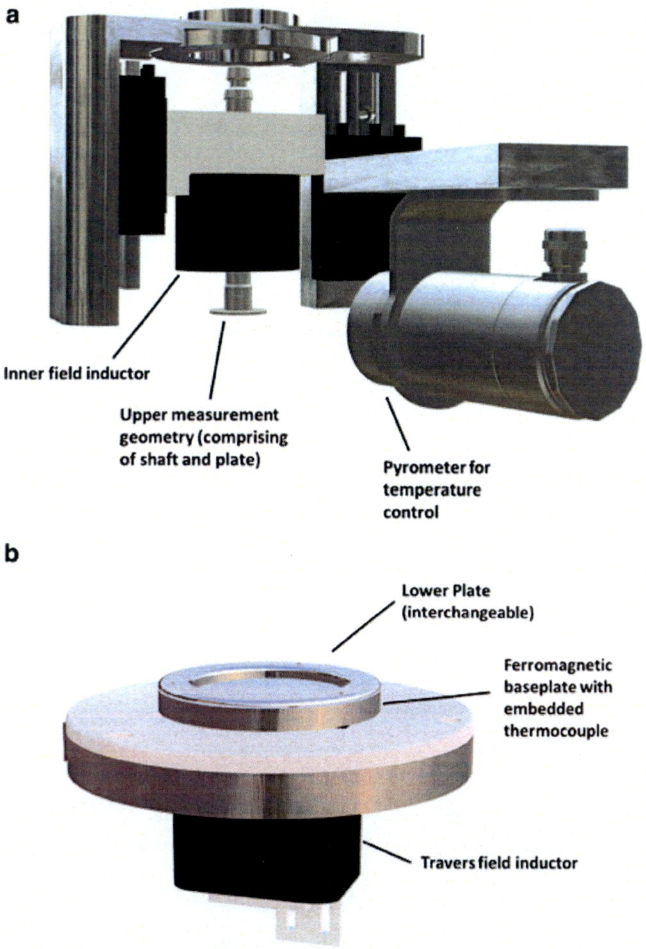

Fig. 1 **a** Upper heating unit—inductor and pyrometer are vertically adjustable for fine alignment. **b** Lower heating unit

adhesives investigated do not incorporate any particles excitable through induction heating. Thus, the development focuses on attaining a rapid temperature increase at the plate surfaces with are in contact to the adhesive. Concerning the use of shaft and plate material (Fig. 2a) non-ferromagnetic steels, ferromagnetic steels and aluminum were compared regarding their heat-up performance. Figure 2b shows the temperature evolution over time of the plate surface being in contact with the adhesive.

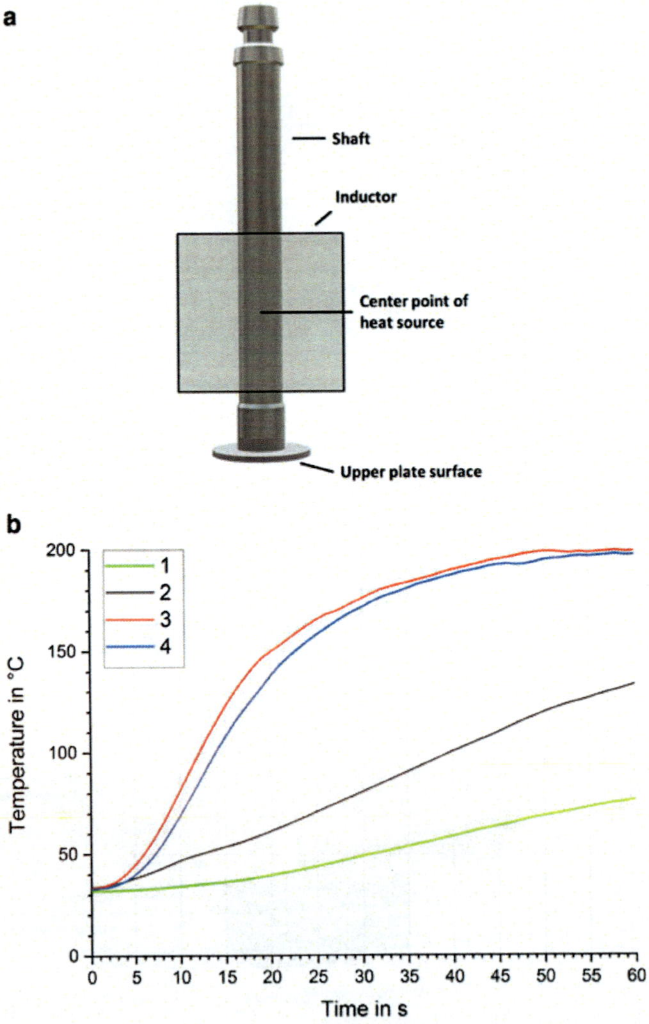

Fig. 2 **a** Measurement geometry with schematic illustration of the inductor. **b** Temperature of upper plate surface—the tested material combinations are described in Table 1

Table 1 Material combinations for temperature performance tests

Test index	Shaft material	Plate material
1	Non-ferromagnetic steels	Non-ferromagnetic steels
2	Ferromagnetic steel	Ferromagnetic steel
3	Ferromagnetic steel	Aluminum
4	Ferromagnetic steel	Aluminum

It is clearly illustrated that the combination of a measurement shaft of ferromagnetic steel and an interchangeable aluminum plate show the fastest temperature evolution reaching 100 °C within 13 s. Hence, this will be used for the prototype. To prevent a heat flux to the motor of the rheometer an air cooling was implemented at the upper end of the shaft. The impact on the upper plate temperature is also shown in the diagram (test index 4)—it is marginally lower than without the cooling. In order to have similar heat flux properties the same ferromagnetic steel was chosen for the baseplate of the lower heating unit which is topped with an interchangeable aluminum plate.

An additional set of pretests was conducted in order to optimize the temperature control process, especially regarding the previously emphasized balance requiring a temperature sensor positioned close to the sample as well as to the heat source. Based on these tests it was decided to use an optical pyrometer positioned just below the inductor for the upper heating unit of the prototype, whereas, a thermocouple embedded in the baseplate controls the lower temperature. The remaining gradient of the measuring position and the plate surface can be calculated through an offset during the heating period. Once the temperature is held constant the gradient decreases to a negligible order. In case of the upper geometry it may be accounted for through a constant offset value.

2.3 *Prototype*

The developed setup is produced and validated with regard to the previously defined requirements. Hereby, a focus is given to two factors with a significant impact on the measurement accuracy: the thermal expansion and the compliance of the upper geometry comprised of the shaft and upper plate (Fig. 2a). Compared to the lower plate and heating unit the upper geometry is much more affected. This becomes apparent when visualizing the temperature distribution along the upper geometry. Figure 3 shows the shaft temperature at the center of the heat source as well as the temperature of the upper plate surface. A difference of several 100 K becomes visible. In order to reach the maximum temperature of 200 °C at the upper plate surface—as specified in the requirements—the heat flux induced in the center point of the inductor creates a temperature increase above 500 °C in the measurement

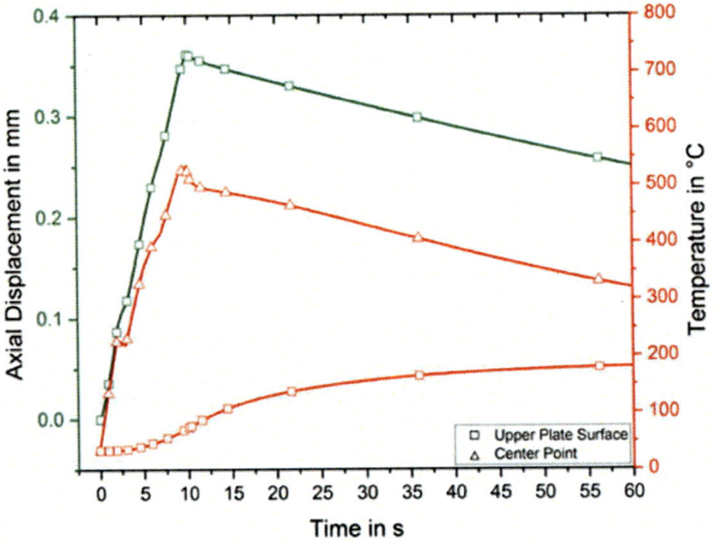

Fig. 3 Thermal expansion of the upper geometry

shaft. After 10 s of heating the induced field decreases while the heat flux spreads along the shaft lowering the temperature gradient between these two points.

2.3.1 Thermal Expansion

The geometry factor from (Eq. 5) incorporates the gap size *h*, assuming it is a constant value along with steady-state temperature conditions during measurement. The thermal expansion in axial direction of the shaft directly impacts the gap size. For steady-state conditions it is compensated for through a constant factor. Knowing the length of the geometry, the expansion factor and the temperature, the change of the gap can be calculated. When measuring rheology prior to a temperature equilibrium this method needs to be revisited. Using the induction heating unit a temperature field along the shaft and upper plate geometry was derived from several heating experiments. It was incorporated into a simulation to determine the displacement of the upper plate (Fig. 3). Knowing the displacement, the change of the gap over time *h(t)* can be calculated. Assuming an initial gap size of 1 mm and a plate diameter of 15 mm this shall be demonstrated for the temperature value of 100 °C during the heating phase (Table 2). The table also shows the change of the geometry factor C_g due to the change of *h*. The geometry factor calculated from the actual values is 27.46% smaller than the factor calculated with constant compensation. Entering the geometry factor into (Eq. 5) it becomes clear that an error in the geometry factor directly causes an error of the same magnitude in the result of the complex shear modulus G^*.

Table 2 Impact of thermal expansion on measurement parameters

	With constant compensation	With actual compensation
Gap size h	0.896 mm	0.65 mm
Geometry factor C_g	$1.803 \times 10^{-4} \frac{1}{mm^3}$	$1.308 \times 10^{-4} \frac{1}{mm^3}$
Torsional compliance J	0,003,309/(Nm)	0,003,398/(Nm)

2.3.2 Compliance

The impact of the torsional compliance on the measurement results is shown by (Eq. 10). For the designed shaft and plate combination the torsional compliance was determined experimentally. The angular deflection θ due to a linearly increasing torque M was monitored at room temperature. Figure 4a shows a corresponding linear increase of the deflection angle. From these results J can be calculated as 0.003309 rad/Nm.

In a second step the results were again incorporated into a simulation. The torque was imposed with the temperature field already used to calculate the thermal expansion. In Fig. 4b the change of the angular deflection over time is plotted for different values of torque. This allows to calculate the variation of the compliance. Again an example is calculated for an upper plate temperature of 100 °C (Table 2). In this case the result of J at actual compensation is 2.69% greater than J calculated with the constant compensation. In order to calculate the modulus data of the sample, J needs to be entered into the set of equations (Eq. 10).

3 Conclusions

The prerequisites for rheological characterizations of adhesive curing reactions under rapid heating conditions have been investigated. The desire to implement fast heating rates leads to novel designed inductive heating units which were integrated into a modern rheometer. The requirements for such a design were described. Key parameters such as the material selection as well as the implementation of the temperature control into the heating units were explored. The control process enables a rapid temperature increase at the upper and lower plate. After reaching a set temperature it can be held within a tolerance of ± 1 K which complies with the standards for curing measurements. Furthermore, the impact of rapid heating on measurement accuracy was evaluated. In order to accommodate a dynamic heating process—in contrast to a state of the art near steady-state temperature increase—an understanding of the underlying mathematical background is necessary. This allows an assessment of the impact of temperature on certain measurement parameters. In case of rapid temperature processes it became visible that the use of constant correction factors for the thermal expansion and the torsional compliance of the upper geometry lead to an error. Especially, when looking at the change of the gap size h, hence the geometry

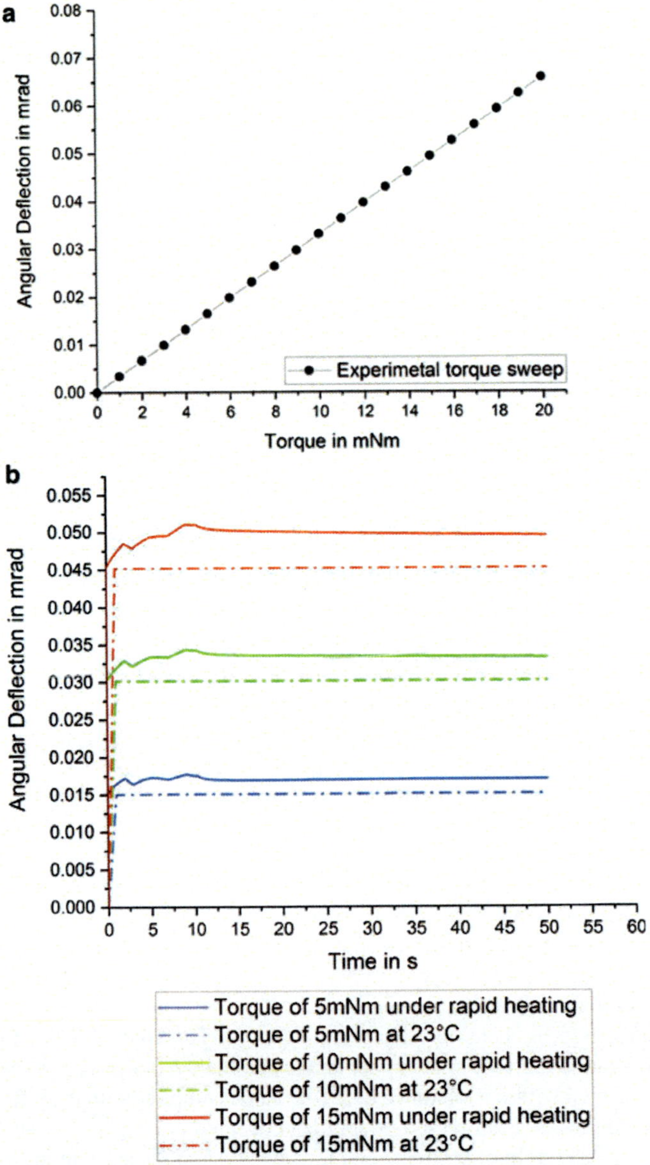

Fig. 4 **a** Compliance of the upper geometry at room temperature. **b** Compliance of the upper geometry during heating

factor C_g and the basic equations for calculating rheological properties it becomes clear that an error in the gap size directly transfers into an error of the same magnitude. The error in torsional compliance also becomes visible when comparing the constant compensation factor with the actual values. However, the impact decreases since the torsional compliance is factored into the denominator when calculating the complex shear modulus. Due to the findings of this study it is recommended to derive the change of the compensation factors throughout the heating process in the first place. In a second step, when calculating the rheological parameters, the constant compensation factors should be replaced by the varying factors gained during the first step. This is necessary to reflect the exact behavior of the parameters under rapid temperature increase. Now, the developed prototype may be used to identify changes in the reaction kinetics of adhesives. When comparing DSC tests performed at low heating rates with those at high heating rates for some adhesives two sectors showing a reaction become visible. How this observation affects the build-up of the modulus shall be investigated in the future.

Acknowledgements The authors thank the German Research Foundation (DFG) for financially supporting this project. Further, we are grateful to Anton Paar for equipment loans and to Amr-Ismail Musa and Omid Zarmandili for performing various pretests.

References

Baehr, H.D., Stephan, K.: Wärme- und Stoffübertragung. Springer, Berlin (2004)

Chaves, F.J.P., Da Silva, L.F.M., de Moura, M.F.S.F., Dillard, D.A., Esteves, V.H.C.: Fracture mechanics tests in adhesively bonded joints: a literature review. J. Adhes. **90**(12), 955–992 (2014)

Cognard, P. (ed.): Handbook of Adhesives and Sealants, Volume 2: General Knowledge, Application of Adhesives, New Curing Techniques. Elsevier, Amsterdam (2006)

Da Silva, L.F.M., Dillard, D.A., Blackman, B., Adams, R.D. (eds.): Testing Adhesive Joints: Best Practices. Wiley-VCH Verlag, Weinheim (2012)

Da Silva, L.F.M., Öchsner, A., Adams, R.D. (eds.): Handbook of Adhesion Technology. Springer, Berlin (2011)

Deer, J.J.: Rheometer. US Patent 4,878,379 (1989)

Ewoldt, R.H., Johnston, M.T., Caretta, L.M.: Experimental challenges of shear rheology: how to avoid bad data. In: Spagnolie, S.E. (ed.) Complex Fluids in Biological Systems, pp. 207–241. Springer, New York (2015)

Fisk, M.: Induction heating. In: Hetnarski, R. B. (ed.) Encyclopedia of Thermal Stresses, pp. 2419–2426. Springer, Dordrecht (2014)

Frauenhofer, M., Kunz, H., Dilger, K.: Fast curing of adhesives in the field of cfrp. J. of Adhes. **88**, 406–417 (2012)

Gottlieb, M., Macosko, C.W.: The effect of instrument compliance on dynamic rheological measurements. Rheol. Acta **21**, 90–94 (1982)

Henkel AG & Co. KGaA: Technical datasheet Teroson PU 1510 (2014)

Jennrich, R., Aydogdu, A.B., Lion, A., Johlitz, M., Glaser, S., Stammen, E., Dilger, K.: Experimental analysis and modelling of temperature and humidity-controlled curing. J. Rubber Res. (2021). https://doi.org/10.1007/s42464-021-00085-9

Jennrich, R., Lion, A., Johlitz, M., Ernst, S., Dilger, K., Stammen, E.: Thermomechanical characterization and modeling of fast-curing polyurethane adhesives. Continuum Mech. Thermodyn. **32**(2), 421–432 (2020)

Kegel, K.: Die Praxis der induktiven Warmbehandlung. Springer, Berlin (2013)

Küpfmüller, K., Mathis, W., Reibiger, A.: Theoretische Elektrotechnik. Springer, Berlin (2013)

Läuger, J., Stettin, H.: Effects of instrument and fluid inertia in oscillatory shear in rotational rheometers. J. Rheol. **60**(3), 393–406 (2016)

Laukkanen, O.-V.: Small-diameter parallel plate rheometry: a simple technique for measuring rheological properties of glass-forming liquids in shear. Rheol. Acta **56**, 661–671 (2017)

Liu, C.-Y., Yao, M., Garritano, R.G., Franck, A.J., Bailly, C.: Instrument compliance effects revisited: linear viscoelastic Measurements. Rheol. Acta **50**, 537–546 (2011)

Lucio, B., de La Fuente, J.L.: Rheological cure characterization of an advanced functional polyurethane. Thermochim. Acta **596**, 6–13 (2014)

Lupi, S., Forzan, M., Aliferov, A.: Induction and Direct Resistance Heating. Springer, Cham (2015)

Mezger, T.G.: Das Rheologie Handbuch: Für Anwender von Rotations- und Oszillations- Rheometern. Vincentz Network GmbH & Co. KG, Hannover (2016)

Muller, R., Gerard, E., Dugand, P., Rempp, P., Gnanou, Y.: Rheological characterization of the gel point: a new Interpretation. Macromolecules **24**(6), 1321–1326 (1991)

Netzsch Gerätebau GmbH: Kinexus series redefining rheometer capabilities. http://www.netzsch.com (2021). Accessed on 17 June 2021

Peter, H.-J.: Handbuch Induktives Löten. Peter, Berlin (2014)

Papadopoulos, E., Ginic-Markovic, M., Clarke, S.: Reaction kinetics of polyurethane formation using a commercial oligomeric diisocyanate resin studied by calorimetric and rheological methods. Macromol. Chem. Phys. **209**(22), 2302–2311 (2008)

Rochery, M., Lam, T.M.: Chemorheology of polyurethane. i. vitrification and gelation studies. J. Polym. Sci. B Polym. Phys. **38**(4), 544–551 (2000)

Sánchez Cebrián, A., Zogg, M., Ermanni, P.: Methodology for optimization of the curing cycle of paste adhesives. Int. J. Adhes. Adhes. **40**, 112–119 (2013)

Schröter, K., Hutcheson, S.A., Shi, X., Mandanici, A., McKenna, G.B.: Dynamic shear modulus of glycerol: Corrections due to instrument compliance. J. Chem. Phys. **125**, 214507 (2006)

Sika Deutschland GmbH: Technical datasheet Sikaflex®-360HC (2011)

VDI.: Verein Deutscher Ingenieure: Design of technical products and systems—model of product design. Beuth, Berlin (2019a)

VDI.: Verein Deutscher Ingenieure: Design of technical products and systems—configuration of individual product design processes. Beuth, Berlin (2019b)

Zimmer, B., Nies, C., Schmitt, C., Possart, W.: Chemistry, polymer dynamics and mechanical properties of a two-part polyurethane elastomer during and after crosslinking. Part I: Dry Conditions. Polymer **115**, 77–95 (2017)

Material Extrusion on Enameled Steel Surfaces

H. Schneider, K. Dilger, and S. Hartwig

Abstract Coating steel with enamel results in hard surfaces with high chemical resistance and due to that to good corrosion protection properties. As a result, enamel surfaces are common applications on bathroom and kitchen equipment. Regarding the corrosion potential by additional metal components in bolted joints and the required drilled holes, in the last few years adhesive bonding on glazed surfaces is becoming more common. However, adhesive bonding requires additional adhesive and curing time and therefore, there is a risk of contamination and longer cycle times. In this context, direct additive manufacturing by material extrusion of thermoplastic polyurethane (TPU) applications on enamel is offering new design opportunities and an approach for implementing hybrid structures that combine hard and brittle enamel on one hand with a rubber-elastic joint part on the other. Aminosilane adhesion promoters are able to link TPU and glass or glazed surfaces. (3-Aminopropyl)trimethoxysilane (APTMS) builds self-assembling multilayers on silicon dioxide. Therefore, the effect of wet-chemical coating processes on the adhesion of a TPU structure built on an enamel via additive manufacturing depends on the variable fraction of SiO_2 in the enamel. In this research, surfaces with an additional amorphous silicon dioxide coating compared to clean and plasma-pretreated enamel with an additive manufactured TPU application, regarding their influence on interface and adherence of an application on polymerized APTMS coatings built by material extrusion on the enamel coating. The results of pull-off adhesion testing based on ASTM D4541 show an increase of adhesion on enamel surfaces with an additional SiO_2-coating and a change of the failure pattern from local failure of cleaned and plasma pretreated to global failure of the SiO_2-pretreated surfaces with higher deformation of the thermoplastic elastomer.

H. Schneider (✉) · K. Dilger · S. Hartwig
Institute of Joining and Welding, Technische Universität Braunschweig, Langer Kamp 8, 38106 Braunschweig, DE, Germany
e-mail: hartwig.schneider@tu-bs.de

K. Dilger
e-mail: k.dilger@tu-bs.de

S. Hartwig
e-mail: s.hartwig@tu-bs.de

L. F. M. da Silva and R. D. Adams (eds.), *6th International Conference on Adhesive Bonding 2021*, Proceedings in Engineering Mechanics,
https://doi.org/10.1007/978-3-030-87668-5_5

Keywords (3-Aminopropyl)trimethoxysilane · APTMS · Thermoplastic polyurethane · TPU · Generative manufacturing · Glass · Glazing · Enamel · Adhesion

1 Introduction

1.1 Motivation

Number of applications for adhesive bonding as well as injection molding and injection stamping on for combining thermoplastics with glass or enamel surfaces is rising. Primary objective are material compounds that utilize advantages of their ingredient materials. Examples for adhesive bonded surface are the magic grip series of Kahla porcelain using silicone for providing nonslip properties on porcelain (see Fig. 1) (Terliesner 2018). Elastic polyurethane bonding of rear and front windows in cars are necessary to obtain vibration resistance and crash properties (Koch 1996). Adhesive bonding of bathroom applications on tiles like towel holder and handlebars instead bolted joints with their higher risk of corrosion is becoming more common (Fig. 2). Semi-finished parts of displays made by adhesive bonding of plastic and glass or by direct injection molding of thermoplastics on glass surfaces are another application of glass-thermoplastic compounds.

1.2 Objective

Figure 3 show the objective of this investigation, the combination of different material properties of glazed surfaces and thermoplastics with advantages of generative manufacturing using adhesion promoters instead of conventional adhesive bonding,

Fig. 1 Magic grip series for nonslip propteries on porcelain surfaces, Kahla

Fig. 2 Bonded towel holder and other bathroom applications, tesa

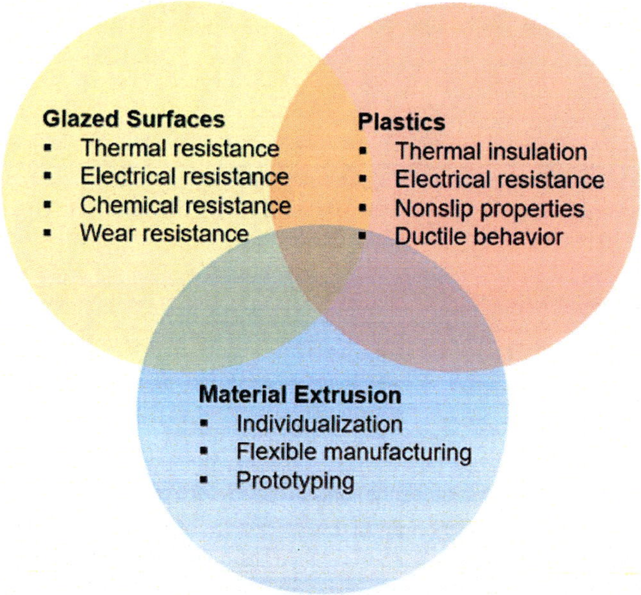

Fig. 3 Combination of different material properties of glazed surfaces and plastics with advantages of material extrusion

injection molding or injection stamping. Therefore suitable materials, pretreatments and adhesion promoters have to be taken in account.

In comparison to adhesive bonding, material extrusion does not need any curing time. Regarding the other two ways for manufacturing injection molding and injection stamping, there is no need for expensive molds in material extrusion. Compared to injection molding, injection stamping is pressure less and therefore similar to material extrusion but takes place on larger surface (Düring and Kosok 1995). Material extrusion is an inexpensive opportunity for low-cost testing of compatibilities of adhesion promoters and joined substrates just as flexible prototyping for trying new designs before construction and manufacturing expensive injection mold. Another field of application would be new repair technologies and generally new ways of direct joining.

Enamel surface consist of fluorspar (CaF_2) and silicates (SiO_2, $MgSiO_4$, $Al_2[O|SiO_4]$) as glazing phase, rutile (TiO_2) for coloring and alkali metals (Na^+, K^+) (Zillig and Sedlbauer 2020; Jebsen-Marwedel 1972; Dietzel 1981).

Possible pretreatments are plasma cleaning and flame treatment. Both of them lead to the removal of organic contaminations by oxidizing. In case of plasma cleaning by atmospheric air plasma jet (AAPJ), ionized air removes organic contamination by oxidization (Lisco et al. 2017). Instead of this, flame pretreatment uses the oxidizing part of a gas flame. Additional burn-up of silane (SiH_4) in this environment could lead to amorphous SiO_2-coatings on the surface (Tokuhashi et al. 1990).

Using aminosilanes like (3-Aminopropyl)trimethoxysilane (APTMS) as adhesion promoters for polyurethane adhesives on glass or glazed surfaces is very common (Golaz et al. 2011) APTMS builds self-assembling multilayers on SiO_2-substrates (Chauhan et al. 2008) by polycondensation (see Fig. 4). Hereafter amino-functional groups of the polymer layer increase the chemisorption to thermoplastic polyurethanes.

2 Experimental Setup

2.1 Substrate

The used substrate is an enamel whiteboard type 4560 from Schäfer with a thickness of 0.52 mm and an enamel thickness of 143.7 ± 5.2 µm measured by inductive coating thickness gauge (type D3, SaluTron). The cutting (H 40 mm × W 40 mm) is carried out with a guillotine shear and afterwards cleaning substrate with acetone for removal of contaminations. Still as-is state after cleaning, enamel shows organic contaminations in XPS-analysis compared to sputtered state that has higher contents of glass phase (Si, Ca, F, O) and coloring (Ti, O) as shown in Table 1.

2.2 Pretreatment and Coating

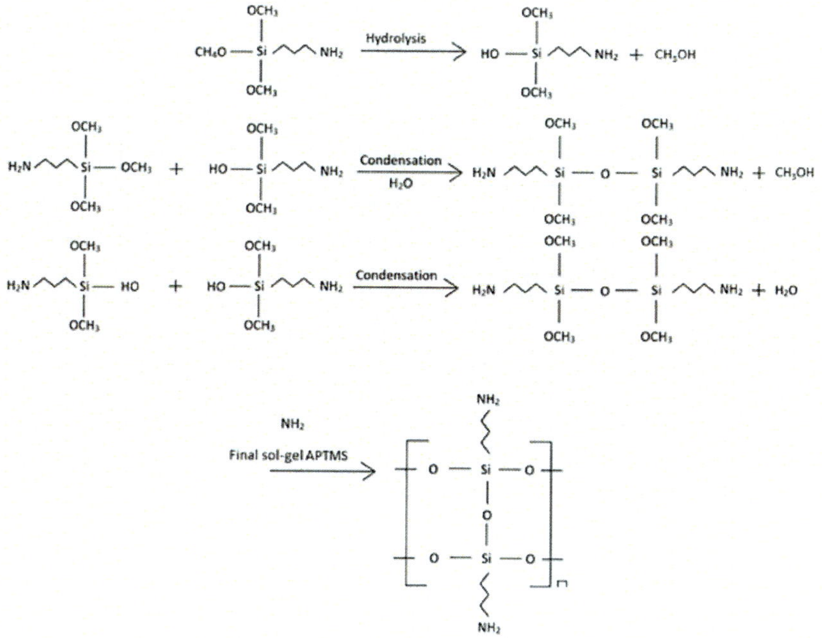

Fig. 4 Polycondensation reaction of APTMS based on (Lecoq et al. 2013)

Table 1 XPS analysis of surface in as-is and sputtered state (Mg Kα, Scofield)

	C 1s	N 1s	O 1s	Na 1s	K 2s	Ca 2p	Si 2p	Ti 2p	F 1s
At%									
As-is/cleaned	46.0	3.1	29.0	1.1	1.4	1.2	14.8	1.0	2.5
Sputtered	23.0	2.2	42.1	2.6	2.3	0.2	20.1	4.5	3.0

Plasma cleaning and SiO_2 coating are mounted on the same linear axis. Static atmospheric air plasma jet from Plasmatreat with a concentrated nozzle (diameter: 4 mm, d = 20 mm, v = 10 mm/s) generator type FG5001 (U = 300 V, PCT = 95%) is used for pretreatment. A Pyrosil repair kit for bonding applications performs for realizing amorphous SiO_2-coating (d = 40 mm, v = 10 mm/s). Oxidizing flame of the Pyrosil hand torch is directed towards the surface in an angle of 45° for preventing it from local overheating and to provide a constant gas flow (Fig. 5).

Before mixture is manually wiped on enamel surface, adding the aqueous coating solution made by 100 ml isopropyl alcohol and 1 ml APTMS with 1 ml H_2O starts polycondensationreaction. After coating, specimens dry at room temperature. Surface free energy calculation by contact angle measurement is conducted following pretreatment and APTMS-coating.

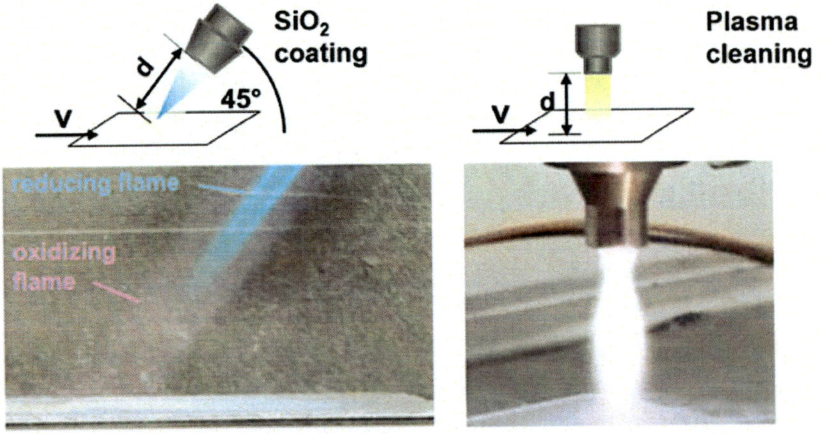

Fig. 5 Overview on silicon dioxide coating (left) and plasma cleaning process as regarded pretreatment techniques

2.3 *Material Extrusion*

An Ultimaker 2+ with nozzle temperature of 235 °C and heat bed temperature of 84 °C is used for manufacturing test specimens; a concave geometry (see Fig. 6) following ASTM D 4541 is generated by material extrusion (Schneider et al. 2021, DIN EN ISO 527 2012). After five layers, the heating bed is switched off. Furthermore, printing a double dolly geometry for determination of tensile strength achievable by means of these printing parameters is necessary (Fig. 7).

The chosen thermoplastic polyurethane of this investigation is a TPU 58 D ($R_{50\%}$ = 16 MPa) from Extrudr. Figure 8 illustrates the influence of the print bed temperature on the print quality. Without cut-off, the high elastic–plastic deformation of the plastic leads to an inaccurate printed image.

Fig. 6 Geometry of printed dolly specimen based on ASTM D 4541

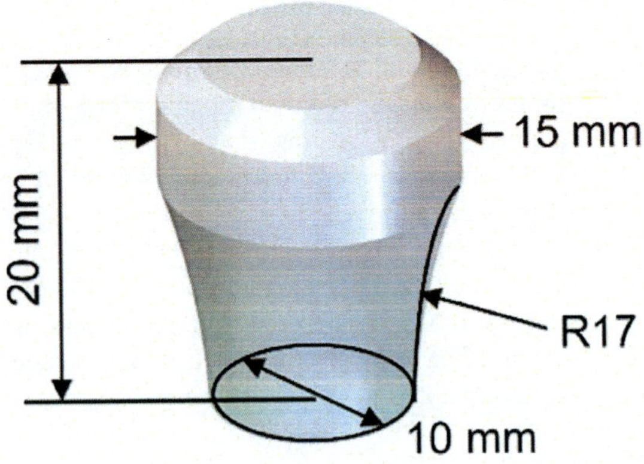

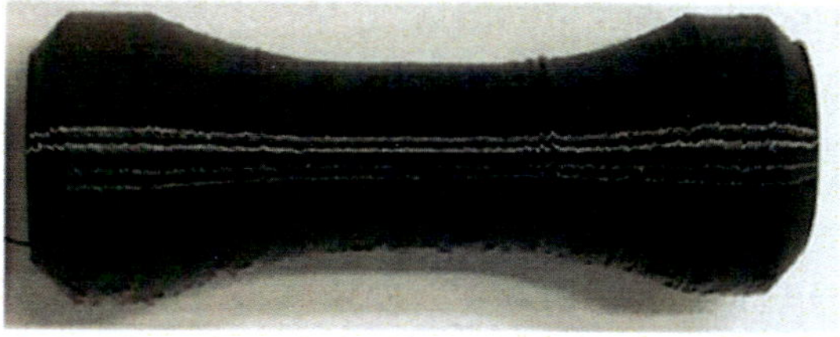

Fig. 7 Double dolly for tensile testing

Fig. 8 Demonstrator for material extrusion on enamel: construction (left), heat bed on until end of process (middle), heat bed off after five layers (right)

2.4 Mechanical Characterization

Before testing in a pull-off adhesion test based on ASTM D4541 at an Instron 10, specimens are backside stiffened by bonding on steel reinforcement using methylmethacrylate (MMA). After testing, proportions of layer delamination and adhesive failure are taken into account using a counting grid. Figure 9 show the setup of pull-off adhesion test.

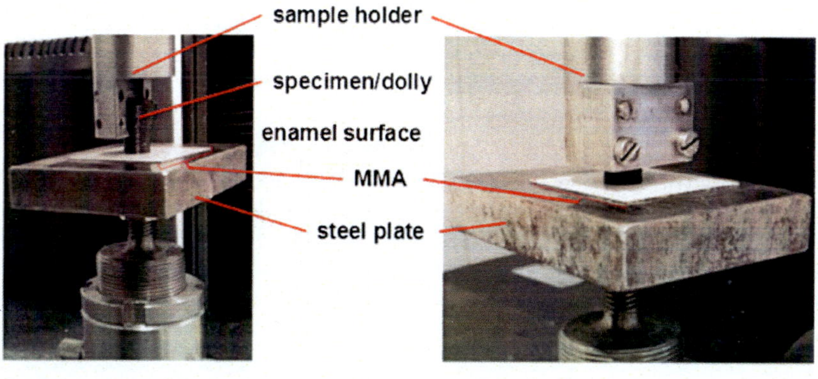

Fig. 9 Pull-off adhesion test setup (v = 10 mm/min)

3 Results and Discussion

Table 2 gives an overview of chemical surface characteristics on pretreated and untreated surfaces and the decrease of organic contamination after pretreatment, where SiO_2-coating shows a stronger influence than atmospheric plasma cleaning. Further, covering by the SiO_2 coating is noticeable by decrease of TiO_2-contents and other elements.

Surface free energy measuring by contact angle and energy determination after Wendt and Owens (1969) only gives qualitative information due to high wettability of plasma cleaned and SiO_2-coated surfaces and surface tension of water (72.8 mJ/m^2 according to Ström et al. 1987). Comparing pretreated surfaces with untreated, dispersion component increases by 40% and polar component by almost 50%. Furthermore, variance of surface free energy decreases from untreated over plasma cleaned to SiO_2-precoated surface (see Fig. 10). Subsequent to APTMS coating, there are almost no differences between pretreated and untreated surfaces due to strong chemical influence of APTMS itself.

The surface in the cleaned state is quite similar to plasma cleaned surface and very inhomogeneous (see Fig. 11a, c). Instead of this, finely dispersed particles connected to surface can be noticed on Pyrosil coatings in Fig. 11e. However, in this comparison resulting APTMS polymer coating in Fig. 11f show the highest homogeneity.

Figure 12 shows the better adhesion properties of plasma cleaned surface (8.0 MPa) compared to untreated surfaces which only rinsed with acetone (6.9 MPa) after APTMS coating Highest adhesion could be recognized on APTMS coatings with SiO_2 coating as pretreatment (11.0 MPa). Comparing reached adhesion and tensile strength of 13.8 MPa for double dolly geometry indicate that the range for further improvements with the same printing parameters is limited.

Compared untreated surface, plasma pretreated surface characterizes higher proportions of layer delamination after pull-off testing. Despite a particularly high adhesion of the TPU dolly to the Pyrosil surface, delaminated layers in contiguous areas where TPU is visible on surface are significantly lower due to higher deformation of the TPU specimen. In addition, variance of estimated layer delamination contents are significantly minor compared with the other two kinds of sample preparation (see Fig. 13).

Further, change of failure pattern from local failure to global failure of the interface is noticeable. Local joint failure on APTMS coatings occurs on acetone cleaned and as well as on plasma pretreated surfaces. It begins with local peeling and ends with local ductile overload fracture of layers (see Fig. 14).

4 Conclusion

Pull-off adhesion tests based on ASTM D 4541 was carried out for thermoplastic polyurethane (TPU) specimens on plasma pretreated, cleaned and SiO_2

Table 2 XPS analysis of surface sputtered, without pretreatment, plasma and SiO_2 pretreated (Mg Kα, Scofield)

	C 1s	Na 1s	K 2s	Mg 2s	Ca 2p	Si 2p	Al 2s	Ti 2p	F 1s	Cl 2p
At%	284 eV	1071 eV	377 eV	90 eV	346 eV	101 eV	111 eV	458 eV	685 eV	200 eV
Sputtered	9.8	3.9	1.2	9.0	1.3	26.6	33.1	6.3	4.3	4.7
No pretreatment	43.4	1.2	3.0	4.7	1.8	18.9	19.3	2.8	3.8	1.1
Plasma cleaned	24.3	4.4	2.0	7.2	2.1	30.2	22.6	2.7	4.1	0.6
SiO_2-coating	15.0	0.4	0.9	6.1	1.1	37.3	37.1	1.0	0.7	0.5

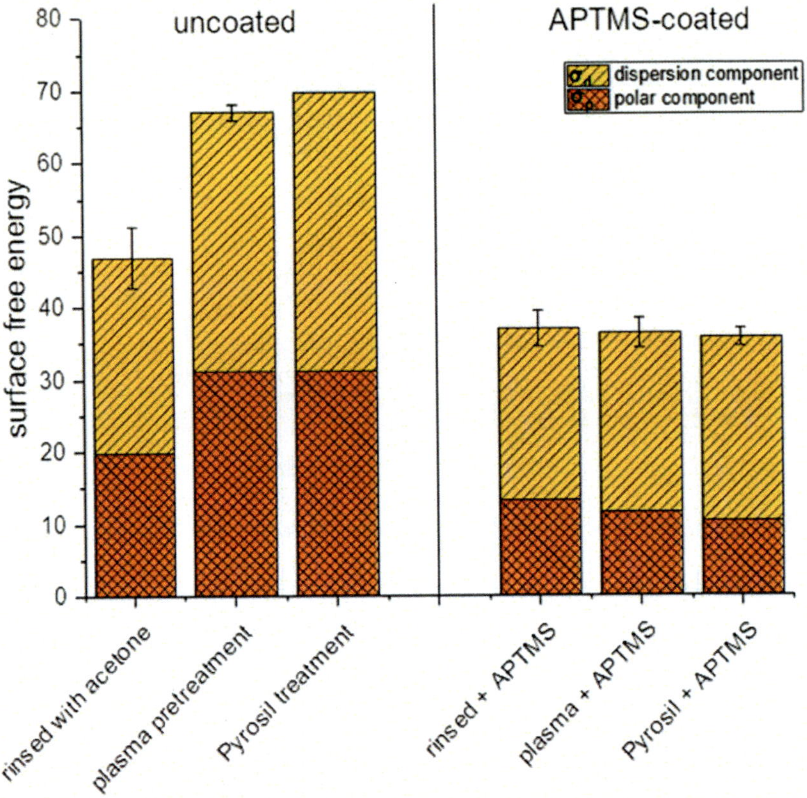

Fig. 10 Surface free energy determined by contact angle measurement after Wendt and Owens 1969 and Ström et al. (1987)

coated enamel surfaces with 3-aminopropyltrimethoxysilane as adhesion promoter. Both pretreated surfaces showed a good wettability at contact angle measurement. Compared to untreated and plasma cleaned state, on amorphous SiO_2 coatings highest homogeneity of assembled APTMS layers was observed. In addition an increase of adhesion occurs from rinsed (6.9 MPa) over plasma cleaned (8.0 MPa) to SiO_2 coated (11.0 MPa) surfaces after APTMS coating was determined. Regarding the printing parameters, a comparison of reached adhesion by SiO_2 coated surfaces with the measured tensile strength of the double dolly geometry of 13.8 MPa shows that the range for further improvements is limited. In addition, a change of failure pattern was determined. Local failure beginning by local delamination and ending by ductile overload fracture of the layers was observed on acetone-cleaned and plasma cleaned surfaces. Instead of this, failure on SiO_2-coated surfaces occurred as global peeling failure with low contents of layer delamination due to higher tensile

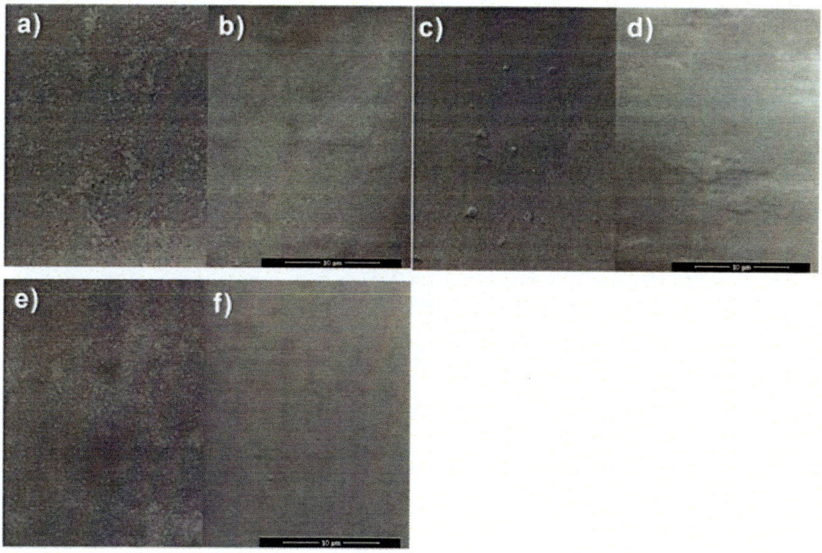

Fig. 11 SEM images of uncoated and coated enamel surfaces (1 kV, mag. 10 k, ETD). **a** unpretreated surface, **b** APTMS-coating without pretreatment, **c** plasma pretreatment, **d** APTMS on plasma pretreated enamel, **e** SiO_2 coating, **f** APTMS-coating on SiO_2 precoating

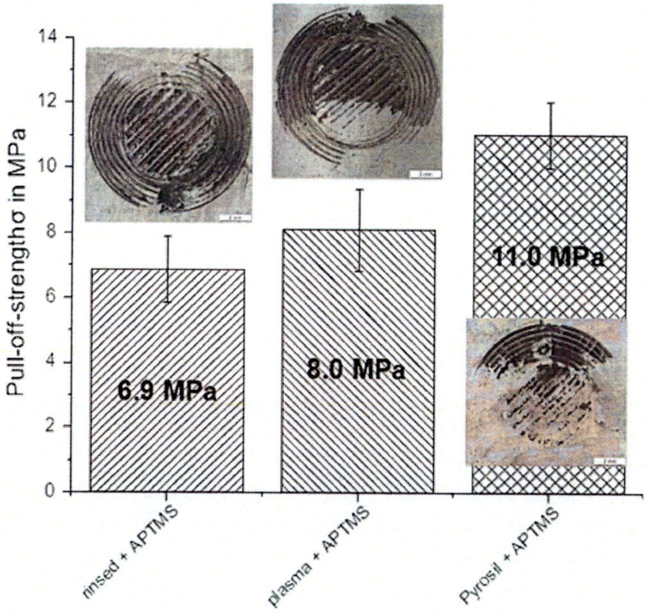

Fig. 12 Comparison of adhesion properties

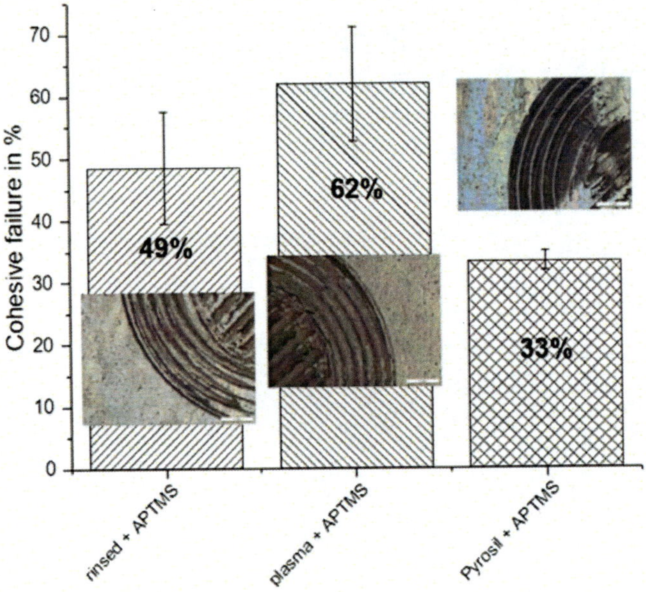

Fig. 13 Failure contents of cleaned, plasma cleaned and SiO_2 coated surfaces after APTMS coating

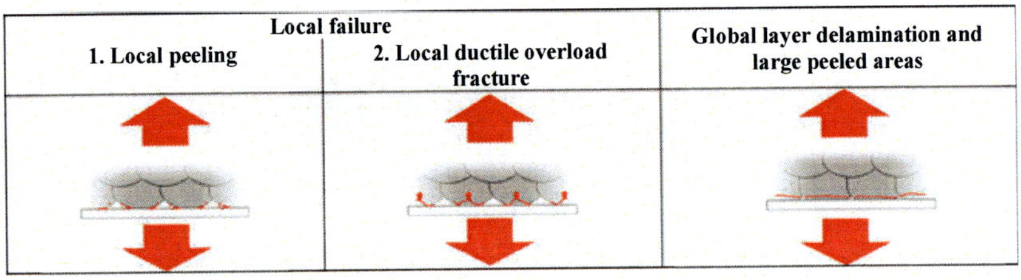

Fig. 14 Local and global failure pattern of extruded TPU 58D on enamel

strain. Further investigations are necessary to clarify damage formation at the interface between enamel and thermoplastic polyurethane. For this reason, microscale computed tomography under static load appears to be a suitable investigation method for a clearer differentiation between local peeling and global elastic deformation under tensile load.

The current investigation considers only mechanical behavior of composite at room climate (23 °C, rh 59%). For this reason, further tests have to be carried out for characterizing aging behavior when exposed to changing climatic conditions, resistance to media (e.g. paraffin oil, tenside) and to characterize mechanical behavior under peel and shear load.

In addition, higher composite strengths are possible by choosing other manufacturing parameters and materials. For example, higher heating bed temperatures could lead to higher adhesion due to lower viscosity of the polymer melt at the interface. Changing TPU to one with significantly higher strength might increase peeling resistance. Moreover, influence of different APTMS contents in wet chemical solution and solvent type on roughness and thus on the mechanical adhesion has to be investigated.

References

Chauhan, A., Aswal, D., Koiry, S., Gupta, S., Yakhmi, J., Sürgers, C., Guerin, D., Lenfant, S., Vuillaume, D..: Self-assembly of the 3-aminopropyltrimethoxysilane multilayers on Si and hysteretic current-voltage characteristics. Appl. Phys. A **90**(3), 581–589 (2008). Doi: https://doi.org/10.1007/s00339-007-4336-7

Dietzel, A.H.: Emaillierung, Wissenschaftliche Grundlagen und Grundzüge der Technologie. Springer, New York (1981). https://doi.org/10.1007/978-3-642-50978-0

DIN EN ISO 527 - Kunststoffe - Bestimmung der Zugeigenschaften - Teil 1: Allgemeine Grundsätze (ISO 527-1:2012); Deutsche Fassung EN ISO 527-1:2012, Deutsches Institut für Normung, Beuth, Berlin (2012)

Düring, J., Kosok, U.: Hinterspritzen von Folienhauben. Kunststoffe **85** (1995)

Golaz, B., Michaud, V., Månson, J.: Adhesion of thermoplastic polyurethane elastomer to galvanized steel. Int. J Adh. Adh. **31**, 805–815 (2011). https://doi.org/10.1016/j.ijadhadh.2011.07.009

Jebsen-Marwedel, H.: Glas, Glasuren, Email und Lack: Gebiete paralleler Forschung von gegenseitiger modellmäßiger Bedeutung. JMT **3**(1), 1–12 (1972)

Koch, S.: Elastisches Kleben im Fahrzeugbau. Dissertation University of Munich (1996)

Lecoq, E., Duday, E. D., Bulou, S., Frache, G., Hilt, F., Maurau, R., Choquet, P.: Plasma polymerization of APTES to elaborate nitrogen containing organosilicon thin films: influence of process parameters and discussion about the growing mechanisms. Plasma Proc. Polym. **10**(3), 250–261 (2013). https://doi.org/10.1002/ppap.201200108

Lisco, F., Shaw, A., Wright, A., Walls, J.M.: Atmospheric-pressure plasma surface activation for solution processed photovoltaic devices. Sol. Ener. **146**, 287–297 (2017). https://doi.org/10.1016/j.solener.2017.02.030

Owens, D., Wendt, R.: Estimation of the surface free energy of polymers. J. Appl. Polym. Sci. **13**, 1741–1747 (1969). https://doi.org/10.1002/app.1969.070130815

Schneider, H., Bollin, A., Dilger, K., Hartwig, S.: Bestimmung der Haftfestigkeit organischer Haftvermittlerschichten mit Hilfe additiv gefertigter Prüfkörper. Tagungsbd. 4. Symp. Mat. 229–241 (2021). https://doi.org/10.21268/20210519-5

Ström, G., Fredriksson, M., Stenius, P.: Contact angles, work of adhesion, and interfacial tensions at a dissolving hydrocarbon surface. J. Colloid Interf. Sci. **119**(2), 352–361 (1987)

Terliesner, S.: Design oder Nicht-sein. Firmenprofil Kahla. Return **4**, 46–49 (2018)

Tokuhashi, K., Horiguchi, S., Urano, y., Iwasaka, M., Ohtani, H., Kondo, S.: Premixed silane-oxygen-nitrogen flames. Comb. Flame **82**, 40–50 (1990).https://doi.org/10.1016/0010-2180(90)90076-4

Zillig, W., Sedlbauer, K.: Werkstoffe – 3. Glas und Keramik. In: Flachdach Atlas Teil D, De Gruyter Oldenburg. 281–290 (2020). https://doi.org/10.1515/9783110496277-029

The Effect of the Viscosity and the Contact Angle on the Pull-Off Strength of Epoxy Resin Modified with Waste Granite Powders

Agnieszka Chowaniec

Abstract It is common practice to modify an epoxy resin coating with additives and fillers. For the sake of sustainable development, in this research, waste granite powders were used as a filler. The most important property of epoxy resin coating is its durability, which is highly affected by the pull-off strength of this coating. The adhesive properties of the epoxy resin coating are also affected by the resin wetting properties. The epoxy resins are characterized by high viscosity which is increase further by the additives. This increase may cause problems when mixing and applying the epoxy resin coating. The influence of the waste granite powders on the viscosity and contact angle of liquid epoxy resin, as well as their influence on the pull-off strength of the hardened epoxy resin coating, is described in this study. Measurements of the contact angle were performed using a test stand of own design and program designed in the MATLAB R2020a environment. The kinematic viscosity measurements were performed using a flow cup. The pull-off strength of the epoxy resin coating was investigated on a concrete substrate using the pull-off method.

Keywords Epoxy resin · Waste granite powder · Viscosity · Contact angle · Pull-off strength

1 Introduction

Based on Analysis of Global Flooring Materials Market (Forecast to 2024, Frost&Sullivan), the segment of coatings constitute 16.4% of the entire European construction chemicals market and a continuous increase in their production is forecasted. Epoxy resin is one type of materials used as coating in civil engineering, the most often as a floor coating or as a protective coating (Francke and Piekarczuk 2020). Epoxy resin coating is the most often used in industrial halls, where high chemical resistance and ease of cleaning are required. However, epoxy resin is synthetic and

A. Chowaniec (✉)
Wroclaw University of Science and Technology, Wroclaw, Poland
e-mail: agnieszka.chowaniec@pwr.edu.pl

L. F. M. da Silva and R. D. Adams (eds.), *6th International Conference on Adhesive Bonding 2021*, Proceedings in Engineering Mechanics,
https://doi.org/10.1007/978-3-030-87668-5_6

its ingredients are harmful to the environment, especially bisphenol A (Xiao et al. 2020). Epoxy resin ingredients are toxic to aquatic organisms (Barboza et al. 2020), are irritating to the skin, and can cause allergic reactions in workers applying this coating (Bangsgaard et al. 2012). Therefore, it is of high importance to reduce the amount of epoxy resin, and even find its natural substitutes. A suitable solution that potentially can reduce the total mass of epoxy resin needed to make the floor coating is to find additives or fillers that will not deteriorate the properties of this coating. Table 1 shows an exemplary studies covering the topic of using epoxy resins with different additives as concrete coatings.

Summarizing the literature review, it can be noted that various additives can improve some of the epoxy resin properties, e.g. pull-off strength, Shore hardness, compressive strength. However, most of the studies present usage of non-waste additives. For the sake of sustainable development, it is worth investigating whether waste materials can be used as additives to epoxy resin.

Considering the positive effect of waste glass powder on pull-off strength (Chowaniec et al. 2020), the interest was drawn to the possibility of using other waste mineral additives such as: fly ash, ground granulated blast furnace slag, silica fume (Czarnecki et al. 2021), limestone powder (De Weerdt et al. 2011), feldspar-quartz (Galińska and Czarnecki 2017) and granite powder (Chajec 2021). In 2018, waste from mining and quarrying accounted for 26.6% of all waste in the EU (Eurostat, Waste statistics 2021). For example, during the mining, cutting and shaping of granite blocks, waste granite powders are produced which are stored in heaps. Fine granite powders are dispersed by the wind, which can pollute the surrounding areas and cause, among others, pneumoconiosis in humans. Incorporating waste granite powders into the hardened epoxy resin coating would recycle this waste and reduce the total mass of the epoxy resin needed to make the coating.

Due to the durability of the epoxy resin coating, as well as the ease of application of liquid epoxy resin, the effect of waste granite powders on the viscosity and contact angle of the liquid epoxy resin and on the pull-off strength of the epoxy resin coating should be investigated. Moreover, for easier application of the epoxy resin and possibility of obtaining a smooth and even surface it is important evaluate the viscosity of the resin and maintain it at low level. Maximum ratio of the addition of powder may depend on shape and diameter of particles, scatter of diameter of particles, interaction between epoxy resin—powder, and powder particles (Stabik et al. 2009). These aspects of the producing the epoxy resins had not to be omitted. For these reasons, it is important to know how the waste granite powders affects the viscosity of liquid epoxy resins. On the other hand the contact angle is one of the most important values measured experimentally during characterization of solids and their wetting characteristics (Drelich et al. 2020). The contact angle is representing the surface free energy and work of adhesion, defining the adhesive properties. The greater the wettability of the material is (the smaller the contact angle), the more work should be done to disconnect the analyzed bodies (Barnat-Hunek 2016). The contact angle may therefore affect the final pull-off strength of hardened epoxy resin coating. The pull-off strength is decisive for the durability of the epoxy resin coating and its average value must not be lower than 1.5 MPa, and single results must not

Table 1 Exemplary studies covering the topic of using epoxy resins with different additive as concrete coatings

Authors	Mixture Content	Analysed properties	Main findings
Chowaniec et al. (2020)	Epoxy resin coating with waste glass powder	Pull-of strength, pores distribution, Na and Si content in subsurface layer	Improvement of pull-off strength and decreasing the volume and size of pores
Guo et al. (2021)	Epoxy resin with TiO_2-graphene	Interfacial bonding strength, water ingression, chloride concentration	The bonding strength was not affect and using up to 0.5% of TiO_2-graphene improved the impermeability
Rousakis et al. (2014)	Epoxy resin enriched with multi-walled carbon nanotubes	Columns behavior during axial compression-decompression-recompression cycles	The small amount of nanotubes remarkably enhanced the mechanical properties of epoxy resin
Sadowski et al. (2021)	Epoxy bonding agent with coconut fibers	Pull-off strength	Increase of pull-off strength was observed
Zheng et al. (2020)	Epoxy resin with graphene oxide	Shore hardness, water contact angle, water absorption	Improvement of shore hardness and decreasing the water absorption rate and chloride concentration
Ying et al. (2021)	Mortar modified with epoxy resin modified with graphene	Compressive strength, chloride concentration and pore volume	Improvement of mechanical properties, higher durability, lower capillary sorptivity and smaller chloride concentration
Shang et al. (2021)	Epoxy resin coating with graphene	Pull-out tests (bonding performance of epoxy-coated steel bars and concrete)	Little effect of graphene on the adhesion bonding performance

(continued)

Table 1 (continued)

Authors	Mixture Content	Analysed properties	Main findings
Krzywiński et al. (2021)	Epoxy resin coating with recycled fine aggregates	The microstructural properties, pull-off strength	Increase in pull-off strength

be lower than 1.0 MPa and performed in 1 per 3 m^2 (Czarnecki et al. 2020). Thus, the epoxy resin coating with waste granite powders must achieve an average pull-off strength of minimum above 1.5 MPa.

The aim of the research is to investigate the effect of waste granite powders on the viscosity and contact angle of liquid epoxy resin and the pull-off strength of epoxy resin coating. Two types of waste granite powders were used in the research, so the aim of the research is also to investigate the influence of the particle size distribution of these powders on the aforementioned properties of the resin. The main aim of the research is to investigate how the viscosity and contact angle of the liquid epoxy resin affect the pull-off strength of the epoxy resin coating modified with waste granite powders.

2 Research Significance

The utilization of waste granite powders in an epoxy resin coating may contribute to reducing the amount of these powders stored in heaps. Adding of fine granite powders into the solid epoxy resin coating will neutralize the negative effects of wind dispersal of these powders. The addition of these powders will reduce the total mass of epoxy resin needed to make the coating. The main novelty of the article is the study of the effect of waste granite powders on the viscosity, contact angle and pull-off strength of the epoxy resin. Moreover the novelty is filling the research gaps on the influence of the particle size distribution of these powders on the aforementioned properties. Another novelty is the study of how the viscosity and contact angle affect the pull-off strength of an epoxy resin coating.

3 Materials and Methods

3.1 Waste Granite Powders

The granite powders used, were waste from the extraction, cutting and processing of granite blocks. Two types of waste granite powders, differing in particle size but having the same chemical composition, were selected for the study. These powders were marked with the symbols P1 and P2. The study of the particle size distribution of the waste granite powders were firstly performed using sieve analysis. Each of the powders were passed through a set of nine sieves with mesh sizes ranging from 2 to 0.025 mm. The mass of the powders remaining on each sieve was weighed, and the particles which had passed through all the sieves, i.e. particles smaller than 0.025 mm, were examined using a scanning electron microscope (model SEM JEOL JSM-6610A by JEOL Ltd.). The particle size distribution of both waste granite powders are shown in Fig. 1. P1 waste granite powder (Fig. 1a) was characterized by a higher content of the finest particles (it contained 18% by weight of particles below 25 μm) when compared to the P2 waste granite powder (Fig. 1b), which contained 10% by weight of particles smaller than 25 μm.

Chemical analysis of waste granite powders were performed using a SEM equipped with energy-dispersive X-ray spectroscopy (EDS). All waste granite powders had the same chemical composition: 76.10% SiO_2, 13.95% Al_2O_3, 0.40% Fe_2O_3, 8.50% $K_2O + Na_2O$, 0.50% MgO, 0.05% CaO and 0.05% TiO_2.

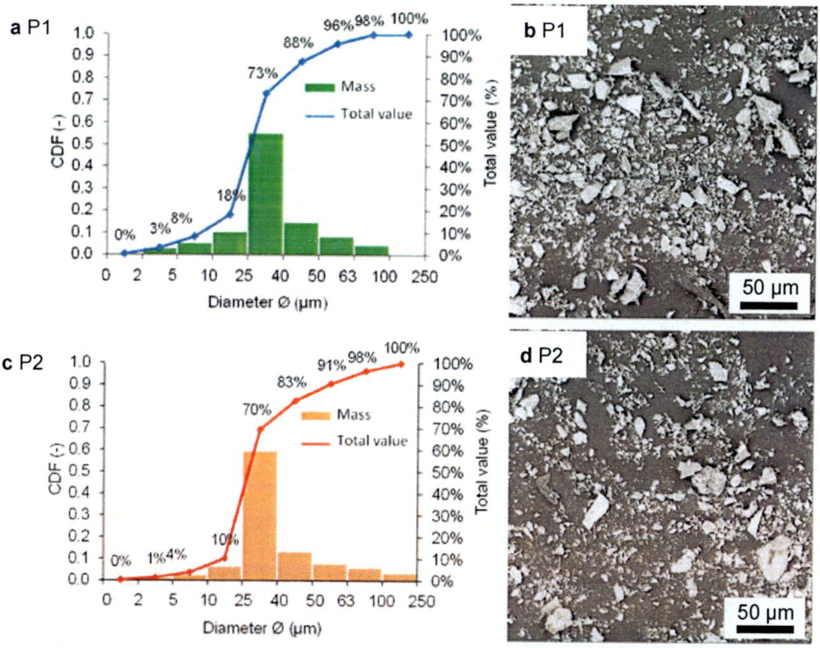

Fig. 1 **a, c** Particle size distribution of P1 and P2 waste granite powder, **b, d** SEM view of P1 and P2 waste granite powder

3.2 *Epoxy Resin and Mixing Proportions of Epoxy Resin with Waste Granite Powders*

The epoxy resin used, was a product intended for use as a coating in industrial floors (Meteor Stone, SI-TECH Sp. z o. o., ul. Dobra 9, 05–306 Jakubów, Poland) and it consisted of two components, A and B, where:

- component A contained: reaction product bisphenol-A-(epichlorhydrin),epoxy resin (number average molecular weight $\leq$ 700); bisphenol-F-epichlorohydrin resin, (number average molecular weight <700); oxirane, mono[(C12-14-alkyloxy)methyl] derivs; benzyl alcohol.
- component B was a phenalkamine hardener.

The mixing proportions of the A and B components were A: B = 100: 50.

Eleven mixtures of epoxy resin with selected waste granite powders were made, and their mixing proportions are shown in Table 2. The symbol 0 denotes the reference mixture without any waste granite powder. Five mixtures were made for each of the two types of waste granite powders. The content of waste granite powders were gradually changed from 7 to 29% of the total mass of the epoxy resin.

The selected waste granite powders were firstly added in the appropriate amount according to Table 2 to component A and mixed very thoroughly. Then, immediately before the tests, component B (hardener) was added and the mixtures were mixed again until uniform consistency and color were obtained.

0 the reference mixture, without any waste granite powder, P1 and P2 the type of waste granite powder.

Table 2 Mixing proportions of epoxy resin with waste granite powders

Marking of mixtures	A weight ratio of A: B: waste granite powder	Components calculated per 100 g of the mixture		
		A (g)	B (g)	Waste granite powder (g)
0	100: 50: 0	67	33	0
P1/7	100: 50: 12	62	31	7
P2/7				
P1/14	100: 50: 24	57	29	14
P2/14				
P1/19	100: 50: 36	54	27	19
P2/19				
P1/24	100: 50: 48	51	25	24
P2/24				
P1/29	100: 50: 60	48	24	29
P2/29				

3.3 Viscosity Measurements

The kinematic viscosity tests were carried out by filling a standard viscosity cup with liquid epoxy resin (Fig. 2) and the time of a known volume of liquid flow passing through an orifice located at the bottom were measured.

For each mixture of the epoxy resin with waste granite powders (according to Table 2), three measurements of viscosity were made. The first measurement was always carried out 5 min after adding the hardener. The viscosity of the epoxy resin was calculated based on the time the epoxy resin flows out in seconds and the parameters of the cup, according to the standard equation (Eq. 1).

$$\nu = 21.78 \cdot T - 306/T \quad (1)$$

where:

ν = kinematic viscosity (cSt),

T = flow time of liquid to the first break in flow (secs).

The flow cup (mtv messtechnik oHG, Erftstadt, Germany) used had an orifice diameter of 8 mm and its other dimensions were in accordance with ISO 2431: 2019 (ISO 2019). The viscosity range that can be measured with this flow cup is 600–2000 cSt, and the flow time should be in the range of 30–100 s. It is important when using a viscosity cup and when retesting epoxy resins that the temperature of the cup and the liquid is maintained, as ambient temperature makes a significant difference to viscosity and thus flow rate. The temperature during the tests should not change by more than 0.5 °C. Viscosity tests were carried out under controlled laboratory

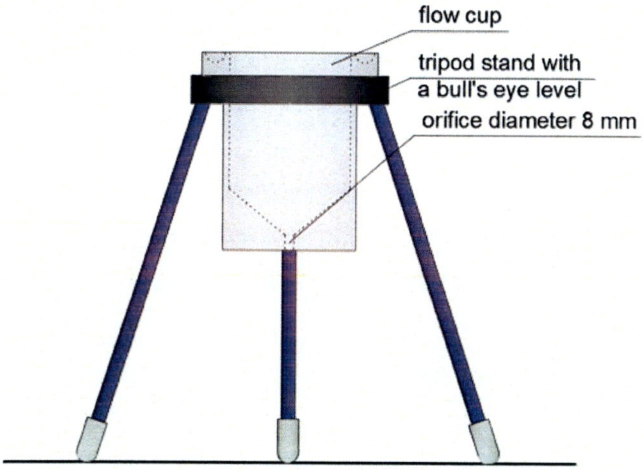

Fig. 2 View of the viscosity test stand

conditions, at an ambient temperature of 25 ± 0.5 °C and a relative air humidity of <65%. The test was carried out in accordance with ISO 2431: 2019.

3.4 Contact Angle Measurements

Measurements of the contact angle were made using a research stand of own design and program created in the MATLAB R2020a environment. The measuring system diagram used for sessile drop imaging and measurement is shown in Fig. 3.

The measurements of the contact angle of epoxy resin with waste granite powders were carried out for the same mixing proportions as for the viscosity measurements (according to Table 2). To eliminate the influence of substrate roughness, contact angle tests were performed on the surface of the microscope slides. The microscope slides were placed on a horizontal measuring table. The syringe with epoxy resin was placed in the syringe holder (Fig. 3). Three droplets were made for each mixing proportions. The first droplet was placed on the surface of the microscope slides after 5 min after adding the hardener. The droplet was illuminated by a source of visible light through a dispersion screen. Each droplet was photographed after 5 s after the drop was deposited. A high resolution camera with single Color Filter Array (CFA) sensor was record profile of epoxy drop with usage of calibrated lens.

The measurements of wetting angle were done by using Axisymmetric Drop Shapc Analysis by Profile (ADSA-P) of sessile drop described by del Rıo and Neumann (1997). Recorded samples were analysed by using the methodology

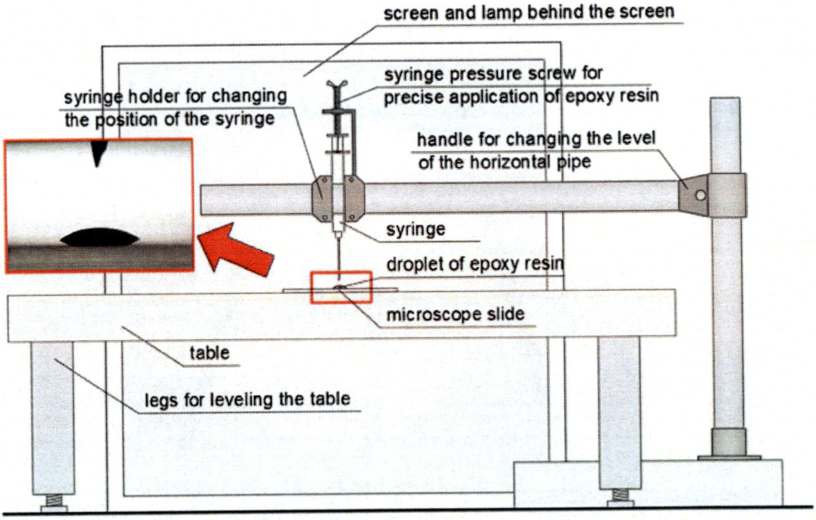

Fig. 3 View of the contact angle test stand

described by Gocławski and Urbaniak-Domagała (2008) which also contains steps for filtering and preprocessing data before using ADSA-P model (del Rıo and Neumann 1997). To avoid spatial measurement noise introduced by image demosaicing (Lukac 2009) data were not demosaiced but normalized by using local histograms (Gómez-Lopera et al. 2000). Pointed above methods of image analysis were implemented in MATLAB R2020a environment.

3.5 *Pull-Off Strength Test*

The pull-off strength tests were performed on the hardened epoxy resin coating on a concrete substrate. A concrete substrate with a thickness of 15 cm and dimensions of 250 cm × 250 cm was made. During concreting, 6 cubes with sides of 15 cm were made to test the compressive strength of the concrete. Compressive strength tests were performed after 28 days in accordance with PN-EN 12390-3. The compressive strength of the substrate concrete was 37.36 MPa. Following the example of (Sadowski et al. 2016), two areas with different methods of substrate surface treatment were separated. The first half of the surface was patched during concreting and in addition, before applying the coating, all loose concrete fragments were removed from the surface with a brush and finally it was thoroughly vacuumed. In this way, a raw surface was obtained. The other half of the substrate surface was mechanically grinded and thoroughly vacuumed prior to coating application. In this way, a ground surface was obtained. Then, measuring squares with dimensions of 20 cm × 18 cm were separated with wooden strips on the entire surface of the concrete substrate. The division of the concrete substrate surface was shown in Fig. 4a. Two rows of measurement squares were intended for the epoxy resin coating with selected waste granite powders. In each row, five measurement squares were separated on both types of surfaces to accommodate the five mixing proportions of epoxy resin with waste granite powder according to Table 2.

According to Fig. 4b, the coating consisted of three layers. The first layer was an epoxy resin-based bonding agent (Meteor Primer, SI-TECH Sp. z o. o., ul. Dobra 9, 05–306 Jakubów, Poland), which consisted of two components mixed in a weight ratio of 100:30. The bonding agent was not modified and was applied to all measuring squares using a paint roller. The thickness of this layer depends on the absorbency of the substrate and is usually 0.1–0.2 mm. The task of the bonding agent is to improve the adhesion of subsequent layers of the coating to the substrate. According to the manufacturer's instructions, the next layer, i.e. epoxy resin coating, was applied 24 h after the application of the bonding agent. As described in Sect. 3.2, the coating used consisted of components: A and B. Waste granite powders were first added to component A in an amount according to Table 2. Component B (hardener) was added immediately before applying the epoxy resin coating to the substrate. According to Fig. 4a, two reference squares (R0 on the raw surface and G0 on the ground surface) were made without any waste granite powder. The epoxy resin coating with a gradually increasing content of waste granite powders was applied to the remaining

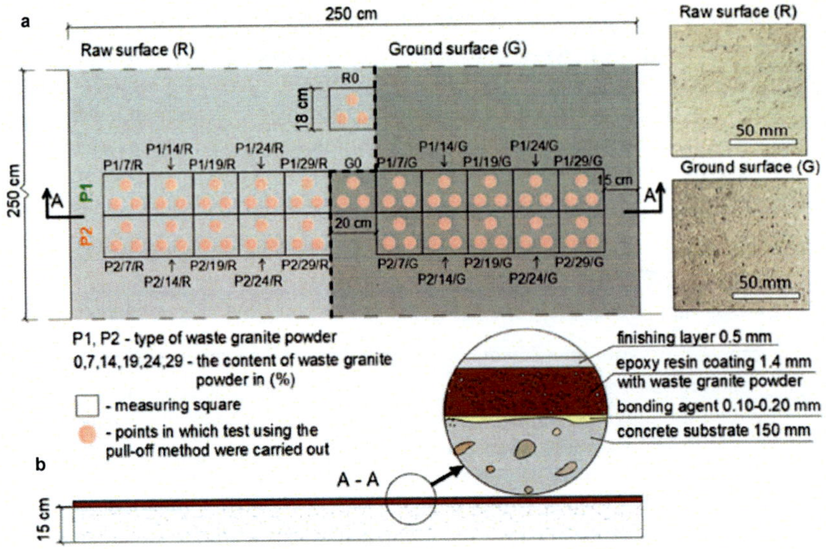

Fig. 4 **a** View of the substrate surface divided into a raw surface and a ground surface, and divided into measurement squares, **b** cross-section through a substrate covered with three layers of coating.

measuring squares. The epoxy resin coating was applied with a notched trowel. The thickness of this layer was 1.4 mm. After another 24 h, a gray finishing layer was applied to all measuring squares using a paint roller, which was also an epoxy based product (Meteor Color). The finishing layer was a two-component product that was mixed in a weight ratio of 100: 15. The task of this layer is to give the coating gloss and the selected color. The thickness of this layer was 0.5 mm. All three layers take 7 days to achieve full mechanical strength. The coating process was carried out under controlled laboratory conditions, at an ambient temperature of 20 ± 2 °C and a relative air humidity of <65%.

The pull-off strength tests were carried out after approximately 3 weeks. The entire measurement procedure was performed in accordance with ASTM D4541 (ASTM 2017). The steps of performing pull-off tests are shown in Fig. 5.

According to ASTM D4541, three measurements are enough to determine the average pull-off strength of the coating, so pull-off tests were performed at three points on each measuring square. Around each measuring point, the entire coating was cut up to a depth of 2–3 mm in the concrete substrate using a core drill (Fig. 5a). The diameter of the drilled area was 5 cm and corresponded to the diameter of the steel disc. The coating surface and the steel disc were thoroughly cleaned with acetone. Acetone does not react with the hardened epoxy resin coating and therefore has no effect on the pull-off strength results. Epoxy glue (Epidian 5 epoxy resin with the Z1 hardener produced by Ciech Sarzyna, Nowa Sarzyna, Poland) mixed with cement was applied to the surface of the coating and to the underside of the steel disk. After the glue was fully cured, pull-off tests were performed using an automatic device

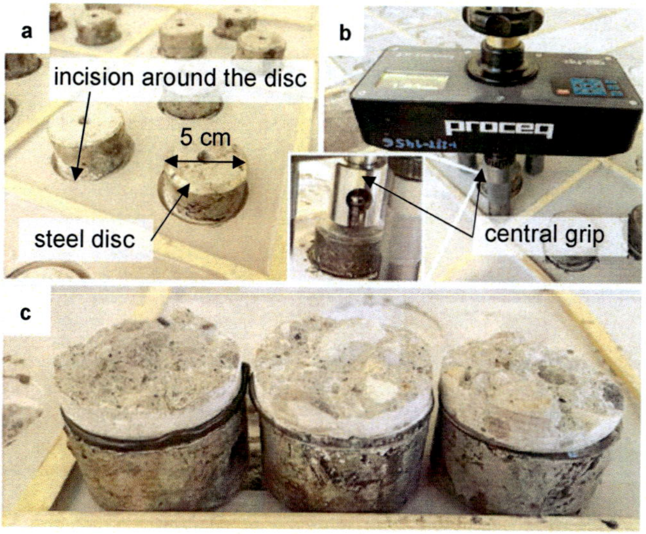

Fig. 5 The steps of the pull-off test: **a** steel discs after sticking, **b** the process of peeling off the steel discs, **c** steel discs after pull-off test

DY-216 (Proceq, Switzerland). The pull-off strength values were calculated from the equation (Eq. 2).

$$X = \frac{4F}{\pi d^2} \tag{2}$$

where:

X = pull-off strength achieved at failure (MPa),
F = maximum force applied to the test surface at failure (N),
d = steel disk diameter (50 mm).

4 Results and Discussion

4.1 Viscosity Measurements Results

Figure 6 shows the test results in the form of a graph of the dependence of the average kinematic viscosity of epoxy resin on the content of waste granite powders.

On the basis of the obtained results, it can be seen that for both waste granite powders very similar results were obtained. As the content of waste granite powders increased, the kinematic viscosity of the liquid epoxy resin increased. The kinematic

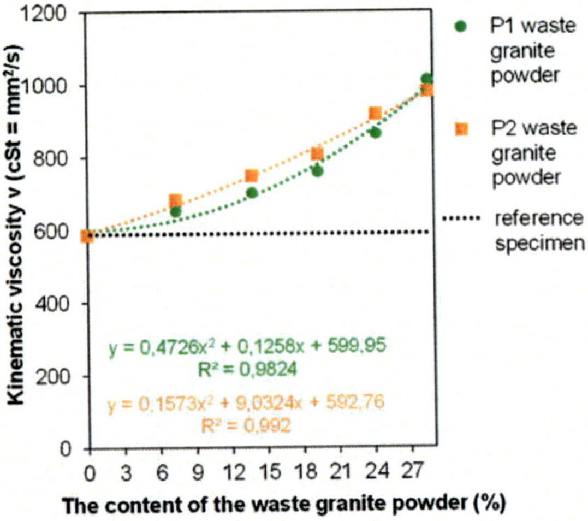

Fig. 6 The relation between the content of the waste granite powders and the kinematic viscosity of epoxy resin

viscosity ranged from 586 mm^2/s (for the reference specimen without any powder) to a maximum of 1009 mm^2/s (for the specimen with 29% P1 powder content). The measurements were very repeatable and the standard deviations were only from 3 to 12 mm^2/s. For each waste granite powders, a trend line was selected in the form of functions, the formulas of which are shown in Fig. 6.

4.2 Contact Angle Measurements Results

Figure 7 shows the test results in the form of a graph of the dependence of the average contact angle of epoxy resin on the content of waste granite powders.

As the content of waste granite powders increased, the contact angle increased from about 30.8° (for epoxy resin without powder) to a maximum of about 37.2° for epoxy resin with P1 powder and to a maximum of about 39.8° for epoxy resin with P2 powder. Standard deviations ranged from 0.7° to 1.8°. For each waste granite powders, a trend line was selected in the form of functions, the formulas of which are shown in Fig. 7.

4.3 Pull-Off Strength Test Results

Tables 3 and 4 show the single pull-off strength results, the average pull-off strength (arithmetic mean AM) and standard deviation (SD) for a coating with P1 waste granite powder (Table 3) and for a coating with a P2 waste granite powder (Table 4).

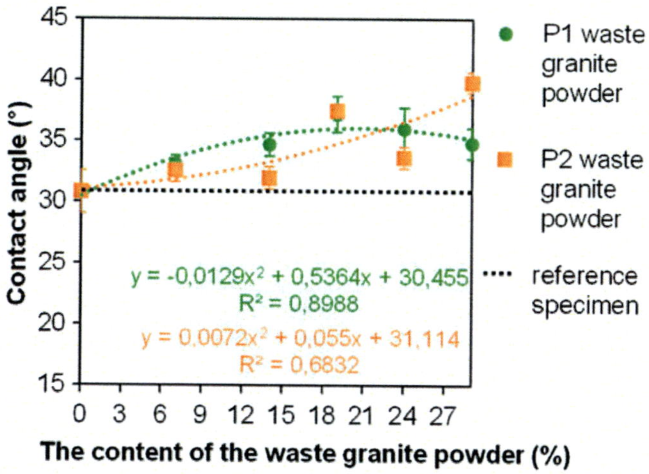

Fig. 7 The relation between the content of the waste granite powders and the contact angle of epoxy resin

The mode of failure was specified for each measurement point in accordance with ASTM D4541.

During the tests, 3 different modes of failure were noticed (Fig. 8): substrate failure A, substrate failure B and failure between the layers of the coating. The substrate failure A is characterized by a detached bulk cement paste and aggregate (Fig. 8a), and the substrate failure B is characterized by a detached bulk cement paste without aggregate (Fig. 8b). These both modes of failure are named cohesive failures in the substrate—in these cases the pull-off strength value mainly depends on the tensile strength of the top layer of the substrate. Failure between the layers of the coating is not the expected type of failure and is due to the lower tensile strength of the coating compared to the tensile strength of concrete. The substrate failure A occurred in almost all measurement points (in 31 out of 36 points), the substrate failure B occurred only in two measurement points, and the failure between the layers of the coating in 3 measurement points.

For both reference squares (R0 on the raw surface and G0 on the ground surface), almost the same average pull-off strength was obtained (2.89 and 2.91 MPa). The average pull-off strength (AM) of the coating with P1 waste granite powder on the raw surface was from 2.54 to 3.09 MPa (average 2.81 MPa), and for the coating on the ground surface from 2.24 to 3.29 MPa (average 2.86 MPa). The average pull-off strength (AM) of the coating with P2 waste granite powder on the raw surface was from 2.09 to 2.89 MPa (average 2.58 MPa), and for the coating on the ground surface from 2.20 to 3.16 MPa (average 2.62 MPa). Based on the above results, it can be concluded that on average higher average pull-off strength values were obtained for a coating with P1 waste granite powder, thus a powder containing more finer particles (Fig. 1). Comparing the average values obtained for both types of surface treatment

Table 3 Summary of the pull-off strength of the coating with P1 waste granite powder

Series	Measuring square	The content of powder (%)	Results (MPa)	AM (MPa)	SD (MPa)	Mode of failure
Raw surface (R)	R0	0%	3.14; 2.73; 2.79	2.89	0.22	All failure A
	P1/7/R	7%	2.38; 2.74; 2.69	2.60	0.20	All failure A
	P1/14/R	14%	2.84; 3.12; 2.05	2.67	0.56	All failure A
	P1/19/R	19%	2.81; 2.82; 3.63	3.09	0.47	1 × failure B, 2 × failure A
	P1/24/R	24%	3.18; 3.05; 3.01	3.08	0.09	All failure A
	P1/29/R	29%	3.52; 2.42; 2.68	2.54	0.13	All failure A
		Average for raw surface		2.81	0.28	
Ground surface (G)	G0	0%	2.55; 3.21; 2.98	2.91	0.34	All failure A
	P1/7/G	7%	2.81; 3.21; 2.73	2.92	0.26	All failure A
	P1/14/G	14%	2.19; 2.19; 2.35	2.24	0.09	All failure A
	P1/19/G	19%	2.76; 2.84; 3.33	2.98	0.31	1 × failure B, 2 × failure A
	P1/24/G	24%	2.60; 2.57; 3.26	2.81	0.39	All failure A
	P1/29/G	29%	2.55; 3.62; 3.69	3.29	0.64	All failure A
	Average for ground surface			2.86	0.34	

R0, G0 the reference mixture, without waste granite powder, *AM* arithmetic mean, SD standard deviation

(raw and ground), higher average values were obtained for the ground surface (2.86 and 2.62 MPa) while compared to the raw surface (2.81 and 2.58 MPa).

Figure 9 shows the graphs of the relationship between the content of waste granite powders and the pull-off strength of the coating. The dark green line represents the results for the epoxy resin coating with P1 waste granite powder on the raw surface, and the light green line on the ground surface. The dark orange line represents the results for the epoxy resin coating with P2 waste granite powder on the raw surface, and the light orange line represents the results for the ground surface.

As can be seen in the Fig. 9, most of the coating's average pull-off strength values were below the average pull-off strength obtained for the reference squares (black

Table 4 Summary of the pull-off strength of the coating with P2 waste granite powder

Series	Measuring square	The content of powder (%)	Results (MPa)	AM (MPa)	SD (MPa)	Mode of failure
Raw surface (R)	R0	0%	3.14; 2.73; 2.79	2.89	0.22	All failure A
	P2/7/R	7%	2.84; 2.17; 2.20	2.40	0.38	All failure A
	P2/14/R	14%	2.52; 2.92; 2.66	2.70	0.20	All failure A
	P2/19/R	19%	2.91; 2.71; 2.37	2.66	0.27	1 × fail. btw. layers, 2 × fail. A
	P2/24/R	24%	2.48; 2.50; 3.31	2.76	0.47	All failure A
	P2/29/R	29%	2.31; 2.01; 1.94	2.09	0.20	All failure A
	Average for raw surface			2.58	0.29	
Ground surface (G)	G0	0%	2.55; 3.21; 2.98	2.91	0.34	All failure A
	P2/7/G	7%	2.74; 2.78; 1.85	2.46	0.53	All failure A
	P2/14/G	14%	2.28; 1.98; 2.33	2.20	0.19	1 × fail. btw. layers, 2 × fail. A
	P2/19/G	19%	2.59; 2.19; 2.33	2.37	0.20	1 × fail. btw. layers, 2 × fail. A
	P2/24/G	24%	2.42; 2.82; 2.72	2.65	0.21	All failure A
	P2/29/G	29%	3.03; 3.40; 3.05	3.16	0.21	All failure A
	Average for ground surface			2.62	0.28	

R0, G0 the reference mixture, without waste granite powder, *AM* arithmetic mean, SD standard deviation

dotted line). However, all the results obtained for the pull-off strength of the coating were above 1.5 MPa, thus above the required minimum value of the pull-off strength of the coating. It should also be noted that a large proportion of the points were within the yellow band, i.e. within the standard deviation of the reference coating. It is not possible to clearly indicate the increasing or decreasing tendency of the influence of waste granite powders on the pull-off strength of the coating.

Fig. 8 Three modes of failure obtained during the tests: **a, b** two cohesive failure modes, **c** failure between the layers of the coating

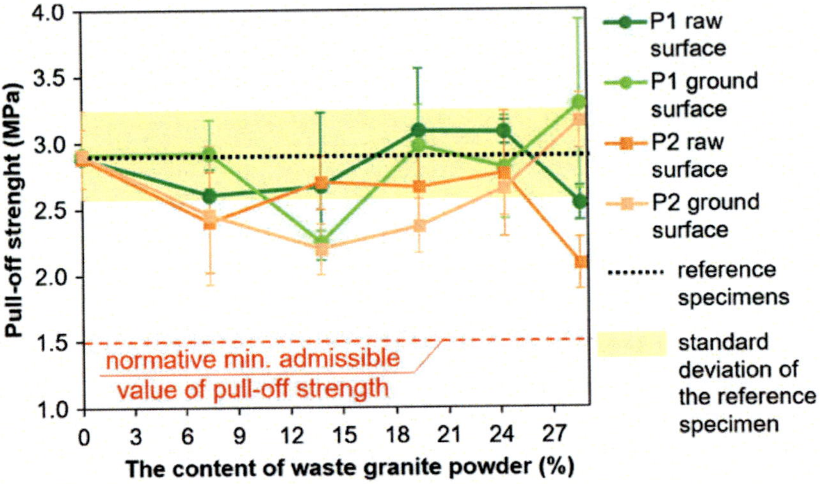

Fig. 9 The relation between the content of the waste granite powders and the pull-off strength of coating

4.4 Dependence of Viscosity and Contact Angle on Pull-Off Strength

In order to investigate the influence of viscosity and contact angle of the liquid epoxy resin on the pull-off strength of epoxy resin coating, the diagrams shown in Fig. 10 were prepared. Plots of the pull-off strength of the coating vs. the viscosity (Fig. 10a) and the contact angle of the epoxy resin (Fig. 10b) were constructed. As can be seen in Fig. 10a, the increase in epoxy resin viscosity caused first a decrease and then an increase in the pull-off strength of the coating to its initial value. Similarly, in Fig. 10b, an increase in the contact angle of the epoxy resin resulted in a decrease and then an increase in the pull-off strength of the coating to the initial value. On this basis, it can be assumed that the addition of waste granite powders in the amount up

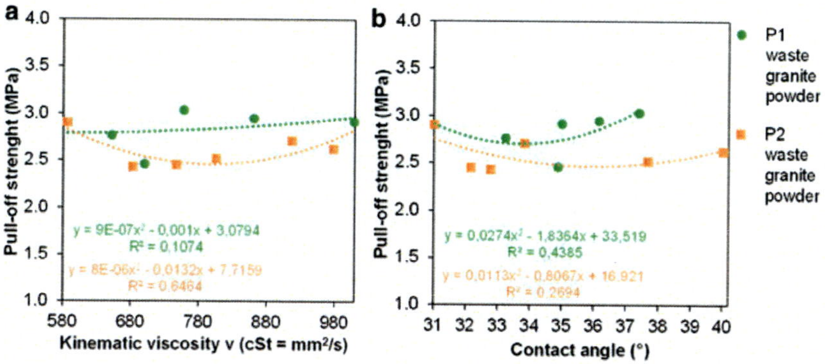

Fig. 10 The relation between the pull-off strength of coating and: **a** kinematic viscosity, **b** contact angle of epoxy resin

to 29% of the epoxy resin mass causes such a negligible increase in the viscosity and contact angle of the epoxy resin that it does not adversely affect the pull-off strength of the epoxy resin coating.

5 Conclusions

The article presents the results of tests of viscosity and contact angle of liquid epoxy resin as well as tests of epoxy resin coatings pull-off strength. The epoxy resin was modified with two types of waste granite powders in the amount of 7–29% of the resin mass. The powders used had the same chemical composition but a different particle size distribution. The tests showed that:

- The addition of waste granite powders increases the viscosity of the epoxy resin. Almost the same results were obtained for both types of powders. Thus, the difference in the particle size distribution of the powders had a negligible effect on the epoxy resin viscosity.
- The addition of waste granite powders increased the contact angle of epoxy resin from about 30.8° (for the resin without powders) to a maximum of 39.8° (for the resin with waste granite powder).
- When analyzing the pull-off strength test results, it was noticed that all the pull-off strength values obtained were above the required minimum value (1.50 MPa). The lowest average value of the pull-off strength of the coating with waste granite powder was 2.09 MPa. The pull-off strength results obtained for the P1 powder (with a higher content of smaller particles) were slightly higher compared to the results obtained for the coating with the addition of the P2 powder. The highest average pull-off strength was obtained for the sample with 29% P1 powder content (3.29 MPa) for the ground surface. This result is 0.38 MPa higher than the value

obtained for the reference sample (2.91 MPa). Comparing the average values obtained for both types of surface treatment (raw and ground), slightly higher average values were obtained for the ground surface compared to the raw surface.

- When analyzing the effect of viscosity and contact angle of the liquid epoxy resin on the pull-off strength of the epoxy resin coating, it was noticed that the increase in viscosity and the increase in contact angle corresponded first to the decrease, and then to the increase in pull-off strength of the coating to about the initial value.

Replacing part of the epoxy resin mass with waste material can lower the cost of the epoxy resin coating and help to recycle the waste granite powders that are currently stored in heaps. In the future, it would be worthwhile to extend the research to the effect of waste granite powders on other properties of the epoxy resin coating, such as, for example, compressive, tensile and flexural strength.

Acknowledgements The authors received funding from the project supported by the National Centre for Research and Development, Poland [grant no. LIDER/35/0130/L-11/19/NCBR/2020 "The use of granite powder waste for the production of selected construction products."

References

Analysis of Global Flooring Materials Market, Forecast to 2024, Frost&Sullivan

ASTM D4541-17.: Standard Test Method for Pull-Off Strength of Coatings Using Portable Adhesion Testers, ASTM International, West Conshohocken, PA (2017)

Bangsgaard, Thyssen, N.J.P., Menné, T., Andersen, K.E., Mortz, C.G., Paulsen, E., Sommerlund, M., Veien, N.K., Laurberg, G., Kaaber, K., Thormann, J., Andersen, B.L., Danielsen, A., Avnstorp, Ch., Kristensen, B., Kristensen, O., Vissing, S., Nielsen, N.H., Johansen, J.D.: Contact allergy to epoxy resin: risk occupations and consequences. Contact Dermatitis **67**, 73–77 (2012). https://doi.org/10.1111/j.1600-0536.2012.02072.x

Barboza, L.G.A., Cunha, S.C., Monteiro, C., Fernandes, J.O., Guilhermino, L.: Bisphenol A and its analogs in muscle and liver of fish from the North East Atlantic Ocean in relation to microplastic contamination. Exposure and risk to human consumers. J. Hazard. Mater. **393**, 122419 (2020). https://doi.org/10.1016/j.jhazmat.2020.122419

Barnat-Hunek, D.: Zwilżalność i kąt zwilżania. In: Barnat-Hunek, D., Swobodna energia powierzchniowa jako czynnik kształtujący skuteczność hydrofobizacji w ochronie konstrukci budowlanych (in Polish), pp. 23–38. Lublin University of Technology, Lublin (2016)

Chajec, A.: Granite powder vs fly ash for the sustainable production of air-cured cementitious mortars. Materials **14**(5), 1208 (2021). https://doi.org/10.3390/ma14051208

Chowaniec, A, Sadowski, Ł., Żak. A.: The chemical and microstructural analysis of the adhesive properties of epoxy resin coatings modified using waste glass powder. Appl. Surf. Sci. **504**, 144373 (2020). https://doi.org/10.1016/j.apsusc.2019.144373

Czarnecki, S., Sadowski, Ł., Hoła, J.: Artificial neural networks for non-destructive identification of the interlayer bonding between repair overlay and concrete substrate. Adv. Eng. Softw. **141**, 102769 (2020).https://doi.org/10.1016/j.advengsoft.2020.102769

Czarnecki, S., Shariq, M., Nikoo, M., Sadowski, Ł.: An intelligent model for the prediction of the compressive strength of cementitious composites with ground granulated blast furnace slag based on ultrasonic pulse velocity measurements. Measurement **172**, 108951 (2021).https://doi.org/10.1016/j.measurement.2020.108951

De Weerdt, K., Kjellsen, K.O., Sellevold, E., Justnes, H.: Synergy between fly ash and limestone powder in ternary cements. Cem. Concr. Compos. **33**, 30–38 (2011). https://doi.org/10.1016/j.cemconcomp.2010.09.006

del Rıo, O.I., Neumann, A.W.: Axisymmetric drop shape analysis: computational methods for the measurement of interfacial properties from the shape and dimensions of pendant and sessile drops. J. Colloid Interface Sci. **196**, 136–147 (1997). https://doi.org/10.1006/jcis.1997.5214

Drelich, J.W., Boinovich, L., Chibowski, E., Volpe, C.D., Hołysz, L., Marmur, A., Siboni, S.: Contact angles: history of over 200 years of open questions. Surf. Innov. **8**, 3–27 (2020). https://doi.org/10.1680/jsuin.19.00007

Eurostat, Waste statistics.: https://ec.europa.eu/eurostat/statistics-explained/index.php?title=Waste_statistics#Total_waste_generation (2021). Accessed on 1 July 2021

Francke, B., Piekarczuk, A.: Experimental investigation of adhesion failure between waterproof coatings and terrace tiles under usage loads. Buildings **10**(3), 59 (2020). https://doi.org/10.3390/buildings10030059

Galińska, A. Czarnecki, S.: The effect of mineral powders derived from industrial wastes on selected mechanical properties of concrete. IOP Conf. Ser.: Mater. Sci. Eng. **245**, 032039 (2017). https://doi.org/10.1088/1757-899X/245/3/032039

Gocławski, J., Urbaniak-Domagała, W.: The measurement of wetting angle by applying an ADSA model of sessile drop on selected textile surfaces. Fibres Text. East. Eur. **16**, 84–88 (2008)

Gómez-Lopera, J.F., Martínez-Aroza, J., Robles-Pérez, A.M., Román-Roldán, R.: An analysis of edge detection by using the jensen-shannon divergence. J. Math. Imaging vis. **13**, 35–56 (2000). https://doi.org/10.1023/A:1008325607354

Guo, S., Luo, H., Tan, Z., Chen, J., Zhang, L., Ren, J.: Impermeability and interfacial bonding strength of TiO_2-graphene modified epoxy resin coated OPC concrete. Prog. Org. Coat. **151**, 106029https://doi.org/10.1016/j.porgcoat.2020.106029

ISO 2431:2019 Paints and varnishes—Determination of flow time by use of flow cups

Krzywiński, K., Sadowski, Ł., Stefaniuk, D., Obrosov, A., Weiß, S.: Engineering and manufacturing technology of green epoxy resin coatings modified with recycled fine aggregates. Int. J. of Precis. Eng. and Manuf.-Green Tech. (2021). https://doi.org/10.1007/s40684-021-00377-w

Lukac, R. (ed.): Single-Sensor Imaging: Methods and Applications for Digital Cameras. CRC Press, Boca Raton, USA (2009)

PN-EN 12390-3:2019-07 Testing hardened concrete—Part 3: Compressive strength of test specimens

Rousakis, T.C., Kouravelou, K.B., Karachalios, T.K.: Effects of carbon nanotube enrichment of epoxy resins on hybrid FRP–FR confinement of concrete. Compos. B Eng. **57**, 210–218 (2014). https://doi.org/10.1016/j.compositesb.2013.09.044

Sadowski, Ł, Czarnecki, S., Hoła, J.: Evaluation of the height 3D roughness parameters of concrete substrate and the adhesion to epoxy resin. Int. J. Adhes. Adhes. **67**, 3–13 (2016). https://doi.org/10.1016/j.ijadhadh.2015.12.019

Sadowski, Ł., Kampa, Ł., Chowaniec, A., Królicka, A., Żak, A., Abdoulpour, H., Vantadori, S.: Enhanced adhesive performance of epoxy resin coating by a novel bonding agent Constr. Build. Mater. **301**, 124078https://doi.org/10.1016/j.conbuildmat.2021.124078

Shang, H., Shao, S., Wang, W.: Bond behavior between graphene modified epoxy coated steel bars and concrete. J. Build. Eng. **42**, 102481 (2021).https://doi.org/10.1016/j.jobe.2021.102481

Stabik, J., Dybowska, A., Szczepanik, M., Suchoń, Ł: Viscosity measurements of epoxy resin filled with ferrite powders. Arch. Mater. Sci. Eng. **38**, 34–40 (2009)

Xiao, C., Wang, L., Zhou, Q., Huang, X.: Hazards of bisphenol A (BPA) exposure: a systematic review of plant toxicology studies J. Hazard. Mater. **384**, 121488 (2020).https://doi.org/10.1016/j.jhazmat.2019.121488

Ying, G., Song, C., Ren, J., Guo, S., Nie, R., Zhang, L.: Mechanical and durability-related performance of graphene/epoxy resin and epoxy resin enhanced OPC mortar Constr. Build. Mater. **282**, 122644https://doi.org/10.1016/j.conbuildmat.2021.122644

Zheng, W., Chen, W.G., Feng, T., Li, W.Q., Liu, X.T., Dong, L.L., Fu, Y.Q.: Enhancing chloride ion penetration resistance into concrete by using graphene oxide reinforced waterborne epoxy coating. Prog. Org. Coat. **138**, 105389 (2020).https://doi.org/10.1016/j.porgcoat.2019.105389

Investigation of the Cleaning Effectiveness of Laser Radiation for Improved Adhesion on Glass Dust Contaminated Glass Fiber Reinforced Polyamide 6

C. J. A. Beier, R. Seewald, H. Dittmar, J. Weiland, A. Schiebahn, P. Jaeschke, L. Overmeyer, and U. Reisgen

Abstract Fiber reinforced polymers (FRP) offer high stiffness and stability properties but are difficult to bond adhesively due to the characteristically low free surface energies of the polymer matrix. In addition, contaminations on the surface act as a barrier and reduce the adhesion between the FRP and the adhesive. Dust is one of the most common surface contaminations and can stem from a variety of sources, especially in industrial environments. To bond FRPs, the surface needs to be clean and contaminations have to be removed from the surface. Laser cleaning represents an innovative method to remove contaminations and simultaneously pre-treat the surface. This work investigates the cleaning effectiveness of 1064 nm laser radiation on a glass fiber reinforced polyamide 6 substrate contaminated with glass

C. J. A. Beier (✉) · R. Seewald · J. Weiland · A. Schiebahn · U. Reisgen
ISF—Welding and Joining Institute, RWTH Aachen University, Pontstr. 49, 52062 Aachen, Germany
e-mail: beier@isf.rwth-aachen.de

R. Seewald
e-mail: seewald@isf.rwth-aachen.de

J. Weiland
e-mail: weiland@isf.rwth-aachen.de

A. Schiebahn
e-mail: schiebahn@isf.rwth-aachen.de

U. Reisgen
e-mail: reisgen@isf.rwth-aachen.de

H. Dittmar · P. Jaeschke · L. Overmeyer
LZH—Laser Zentrum Hannover e.V., Hollerithallee 8, 30149 Hannover, Germany
e-mail: h.dittmar@lzh.de

P. Jaeschke
e-mail: p.jaeschke@lzh.de

L. Overmeyer
e-mail: l.overmeyer@lzh.de

L. F. M. da Silva and R. D. Adams (eds.), *6th International Conference on Adhesive Bonding 2021*, Proceedings in Engineering Mechanics,
https://doi.org/10.1007/978-3-030-87668-5_7

dust. The investigations are based on SEM (scanning electron microscopy) investigations, surface energy investigations by contact angle measurement, tensile shear tests, and roller peel tests. The treatment of uncontaminated and contaminated samples is compared. Additionally, for the cleaning process the number of transits was varied. The best parameter for both cleaning and pre-treatment was found to be at a pulse fluence of 0.88 J/cm^2.

Keywords Laser · Surface pre-treatment · Polyamide 6 · Surface cleaning

1 Introduction

Application and demand of fiber reinforced plastics (FRP) are increasing every year due to the exceptional lightweight properties of FRP. Especially in the automobile industry, FRPs are utilized to improve the fuel efficiency or battery range by decreasing the car's weight. For FRP, adhesive bonding is a well-established joining process that maintains the unique properties of the FRP and does not damage the compound. However, due to the characteristically low free surface energy of polymers, good adhesion to the matrix requires a surface pre-treatment. Additionally, contaminations on the product's surface can act as a barrier and further weaken adhesion. Therefore, contaminations have to be removed before bonding. Laser radiation can combine both processes, cleaning and surface pre-treatment (Mittal and Lei 2018).

Additionally, a laser process can be observed via breakdown spectrometry. Dittmar et al. (2021) investigated the different spectrometers to detect release agent on CFRP substrates by laser induced breakdown spectrometry.

Weiland et al. (2020) reported good results for laser treatment of glass fiber reinforced polyamide 6. Wavelengths of 355 and 1064 nm at varying laser powers were investigated. The authors achieved a significant increase in bonding strength and almost doubled the lap shear strength with the best parameters. The failure patterns showed parts of cohesive failure.

Napadlek (2009) differentiates between laser cleaning in dry and wet environments. Furthermore, in dry environments Napadlek defined two extreme cases depending on the optical properties of contamination and substrate, as shown in Fig. 1. In the first case, the particle holds low absorption for the laser radiation and the substrate holds high absorption. Therefore, the laser beam does not affect the particle directly, but stimulates the substrate. By heat expansion and ablation of the substrate, the particle is detached from the surface. In the second case, the particle absorbs the radiation, while the substrate holds low absorption and is not affected by the laser beam. The particle thermally expands due to the absorbed energy or is directly ablated.

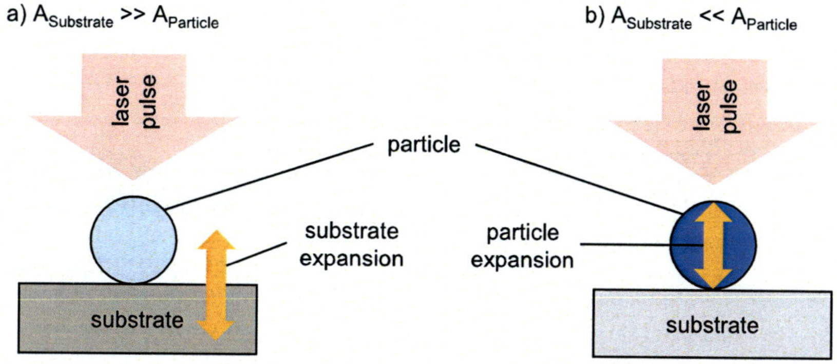

Fig. 1 Extreme cases in dry laser cleaning according to Napadlek (2009)

Mittal and Lei (2018) defined the main process parameters for a dry laser cleaning process to be process and contamination related. Especially the laser fluence influences the cleaning process, with higher fluence resulting in better cleaning effectiveness. In terms of the contamination, besides its optical properties, the particle size is of importance. With increasing particle size, the removal threshold decreases. While the adhesion forces typically are linearly proportional to the particle size, the induced cleaning forces are in general proportional to the square of the particle size. Therefore, the removal of smaller particles becomes more ambitious.

Fourier et al. (2001) investigated the laser cleaning process for dusts on polymer surfaces. The authors combined absorbing dust with absorbing and not absorbing substrates, as well as not absorbing dust on an absorbing substrate. The investigations showed a reduction of the contaminations for all combinations. Best results were achieved with the not absorbing dust with a cleaning efficiency of 95 %.

Kreling et al. (2013) and Rauh et al. (2018) investigated the laser cleaning of release agents on carbon fiber reinforced epoxy. Both authors achieved an increase in lap shear strengths with the laser cleaning process. Best results occurred when a layer of the epoxy matrix was ablated, without uncovering the carbon fibers.

This paper presents an investigation of the laser cleaning effectiveness for glass dust on a glass fiber reinforced polyamide 6 substrate. Optimal laser parameters are identified in regard to the contamination and compared to the pre-treatment process on uncontaminated substrates.

2 Materials and Methodology

The investigations are conducted on an injection molded polyamide 6 substrate with 30 % glass fiber reinforcement (PA6-GF30). The glass fibers have an average fiber diameter of 10 μm and a standard cut length of 3.8 mm. The polyamide is compounded with various additives, giving it a black color. Due to the additives, the substrate absorbs 95 % of radiation in the relevant range of wavelengths (Weiland et al. 2020). Previous investigations showed that the polyamide itself does not absorb a significant share of the radiation for wavelengths between 350 and 2000 nm (Weiland et al. 2020).

To simulate the most common surface contamination, dust, S32 glass bubbles by 3M Deutschland GmbH from Neuss, Germany, were used. These glass bubbles, commonly used as a filler material with low density, have an average diameter of 40 μm with 80 vol.-% of the dust containing particles with diameters between 20 and 75 μm (3M Deutschland GmbH 2017). This distribution of particle sizes represents a wide spectrum of particulate matter in an industrial environment, such as industrial dust or pollen (Möller 2003). The glass bubbles are applied on the substrate in a suspension with isopropanol. This application ensures a homogenous particle distribution on the substrate's surface.

The optical properties of the glass dust is investigated by UV–Vis spectroscopy. The spectrophotometer measures the reflection R and transmission T of light through a substrate. The absorption A is calculated subsequently according to the following formula (Mittal and Lei 2018):

$$A + T + R = 1$$

The result of the laser cleaning process is evaluated optically, analytically and mechanically. The surfaces are investigated via SEM (scanning electron microscopy) to optically determine the particles on the surface. Furthermore, the surface roughness is measured by laser scanning microscopy. The surface roughness is measured in 40 vertical and horizontal lines each, covering an area of 1.5 mm × 1.5 mm. The contact angle of the treated surface is measured to determine remains of the contamination and the effect of the laser pre-treatment on the surface. These measurements are carried out on the basis of DIN EN 828, the surface energy is calculated according to the OWRK method (Kaelble 1970; Rabel 1971), with distilled water, ethylene glycol and diiodmethane used as testing liquids.

The effect of the cleaning process on the bonding strength is mechanically evaluated in roller peel tests and tensile lap shear tests. The roller peel test is conducted according to DIN EN 1464, using 3M Aluminum Foil Tape 425 by 3M Deutschland GmbH, Neuss, Germany. The tape is applied on the sample and loaded with a weight of 10 kg for 72 h before the test is carried out. A Retro Line Shear tester by ZwickRoell GmbH & Co. KG from Ulm, Germany, is used to apply the force at 100 mm/min.

The tensile lap shear tests are carried out according to DIN EN 1465. The samples are bonded in a single lap joint using 3M SW DP 609, a two-component polyurethane-based adhesive by 3M Deutschland GmbH, Neuss, Germany. A constant adhesive layer thickness of 0.5 mm is achieved using glass balls. The test is carried out at a speed of 1.5 mm/min. Due to the number of parameters, three tests are carried out per parameter for statistical purposes. After testing, the failure patterns are analyzed

Table 1 Conversion of laser power input to pulse fluence

Nominal power [%]	Output power [W]	Pulse fluence [J/cm^2]
10	0.53	0.173
20	1.05	0.343
30	2.70	0.882
40	4.57	1.492
50	6.44	2.103
60	8.47	2.765
70	10.36	3.382
80	12.37	4.039

Table 2 Test plan

Contamination	Contamination quantity (ml)	Laser power (%)	Repetitions
Uncontaminated	–	0–50	1
Glass dust	0.5	0–50	1
Glass dust	0.5	0–50	3

according to DIN EN ISO 10365 to obtain more information about the cleaning effectiveness and adhesion.

The laser cleaning process is conducted with a CL20 Laser by Clean-Lasersysteme GmbH from Herzogenrath, Germany. This laser is a pulsed Nd:YAG laser with 1064 nm wavelength and 65 μm spot diameter, operating at 100 ns pulse length. The conversion from nominal power of the laser, which is used to present the results, to output pulse fluence is shown in Table 1. The surface is scanned with a scanning speed of 3000 mm/s and a pulse overlap of 50 % in both scan direction and hatch direction. The laser power and number of repetitions are varied.

The test plan is displayed in Table 2. The glass dust was investigated in a quantity of 0.5 ml of dust suspension with varying the laser power and number of repetitions with 0 % laser power representing no treatment. The laser cleaning is additionally compared to uncontaminated samples that are pre-treated with single transit parameters. The uncontaminated samples are only cleaned with isopropanol.

3 Results

To measure the absorption of the contamination, the glass dust has to be applied on a glass substrate with a clear tape. The glass substrate and tape may influence the measurement and are also analyzed by themselves and in combination. To eliminate the influence of tape and glass substrate, the absorption is calculated for the tape and glass substrate, and the glass dust on tape and glass substrate. Subsequently, the

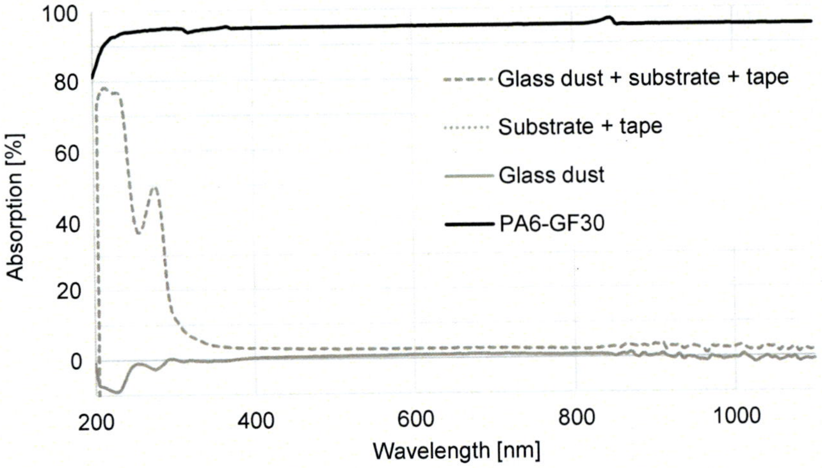

Fig. 2 Absorption spectra of release agent and glass dust

absorption of the tape and substrate are subtracted from the absorption of the glass dust, resulting in the curves shown in Fig. 2. The graph shows a negative absorption coefficient at small and big wavelengths, which is a result of the measurement and calculation. The absorption coefficient can be assumed to be close to 0 in these areas. The signal change at 860 nm is system related. The glass dust does not have a significant level of absorption. This finding shows, that the laser cleaning can only be achieved indirectly by impacting the substrate, not the glass dust itself. Therefore, the laser cleaning process will also act as a pre-treatment process, a differentiation of the cleaning process is not possible. The PA6-GF30 shows a high absorption of over 95% for the majority of the spectrum.

The SEM investigations show a distinct removal of particles. A comparison of an untreated (0 J/cm^2) and a treated sample (2.1 J/cm^2) are shown in Fig. 3. The glass dust particles are visible as white, round particles, the white strips are glass fibers close to the surface. The number of particles on a surface area of 4.7 mm^2 are counted for each sample to evaluate the cleaning effectiveness in a first step. Small and big particles are not distinguished. The resulting graph is displayed in Fig. 3. The number of particles decreases quickly with increasing pulse fluence up to a pulse fluence of 0.88 J/cm^2. For higher fluences the number of particles left on the surface does not decrease significantly.

The measured surface roughness Rz for the contaminated and uncontaminated samples before and after laser cleaning with different laser powers is shown in Fig. 4. The surface roughness is increased by the contamination (0 % laser power). The laser treatment process raises the surface roughness between 15 and 22 μm for a laser power up to 40 %. For 50 % laser power, the surface roughness increases to almost 35 μm for both single transit samples. For three transits, the roughness increases only slightly to 24.34 μm. The cleaning process cannot be observed in the surface

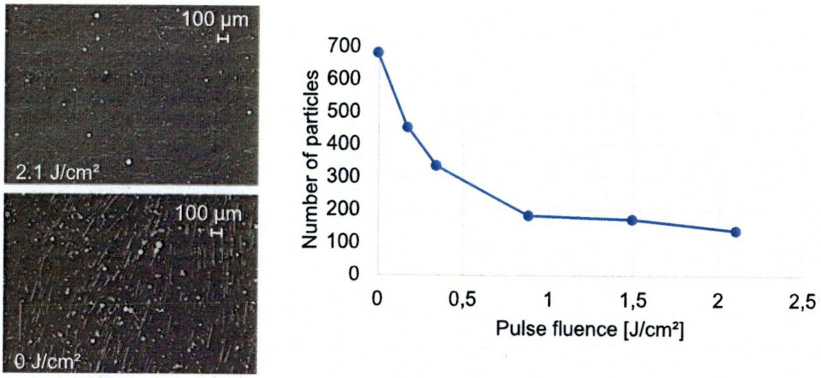

Fig. 3 SEM investigations for first evaluation of cleaning effectiveness for glass dust and single transit

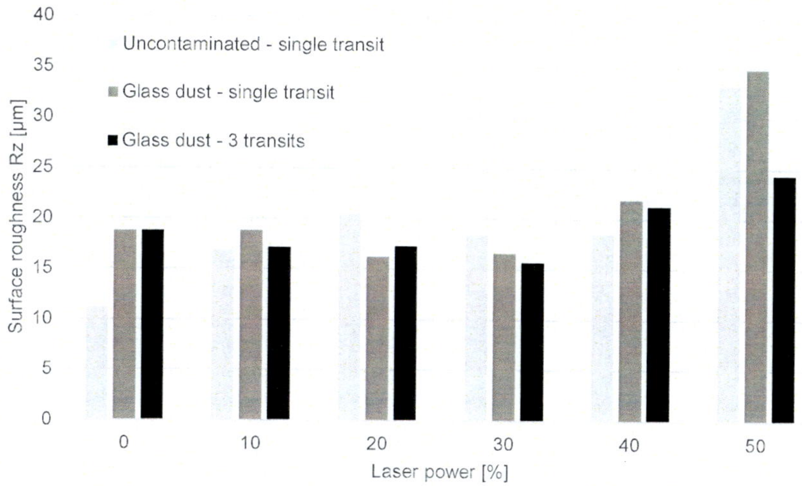

Fig. 4 Surface roughness Rz

roughness. Although the surface roughness decreases slightly for the contaminated samples at laser powers up to 30%, the change is very small and not significant enough to make a statement. For higher laser powers the surface roughness increases due to ablation of the substrate. The laser process is also influencing the roughness for uncontaminated samples by ablation of the substrate.

The measurements of the surface energy are displayed in Fig. 5 and show no significant difference between contaminated and uncontaminated surfaces before treatment (0 %). After treatment, the surface energy decreases slightly for all parameter sets with increasing laser power. The values of contaminated and uncontaminated

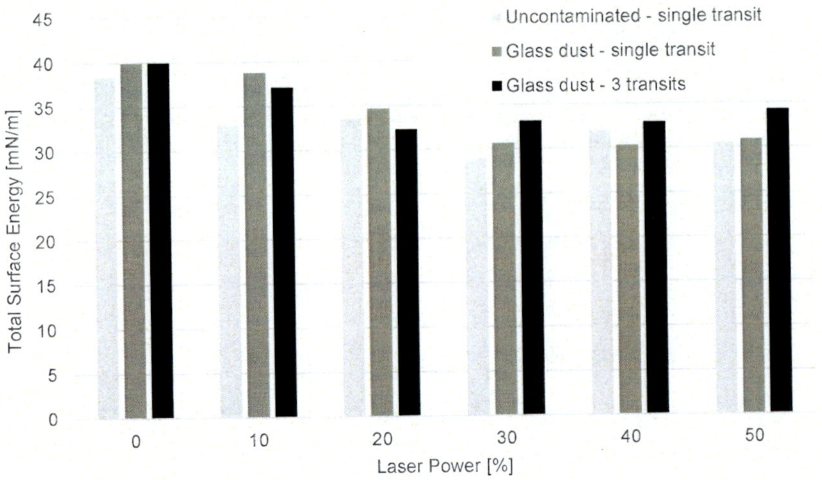

Fig. 5 Surface energy

samples do not differ significantly, besides at 10% laser power, where the surface energy for the uncontaminated sample is lower than for the contaminated ones. This observation indicates that the glass dust acts as a barrier on the surface, reducing the interaction of the laser beam with the substrate. A significant difference between single and triple transit cannot be observed.

The peel resistance, displayed in Fig. 6, shows a substantial difference due to the contamination. Without treatment, the glass dust almost bisects the peel resistance

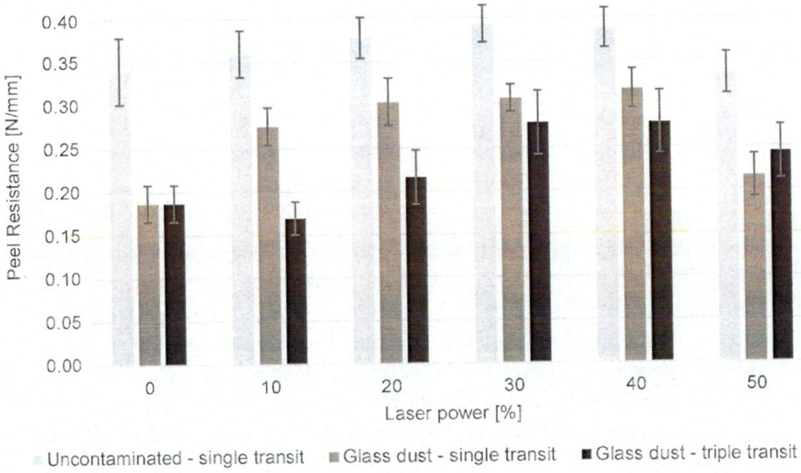

Fig. 6 Peel resistance, error indicators show standard deviation

from 0.34 to 0.19 N/mm. The laser treatment slightly increases the peel resistance for uncontaminated surfaces to a maximum of 0.39 N/mm at 30% laser power. For higher laser powers the peel resistance drops, which can be linked to the increasing surface roughness. The increased roughness leads to decreasing bonding area for the thin adhesive film on the tape. The adhesive film cannot reach and wet the complete roughened surface, therefore the actual bonding area decreases which leads to lower peel resistance. The laser treatment improves the peel resistance for the contaminated samples significantly. Single transit treatment results in higher values. The peel resistance of uncontaminated samples is not reached though. The graphs have a maximum around 30–40% laser power. A laser power of 50% decreases the peel resistance again.

Figure 7 displays the tensile shear strength. Additionally, failure patterns are marked above the bars for patterns other than pure adhesive failure. The tensile shear strength is decreased by the contamination by about 25 %. Without treatment, contaminated and uncontaminated samples fail adhesively (AF). For the uncontaminated samples, the laser treatment doubles the tensile shear strength to a maximum of over 8 MPa at 30 % laser power, with a substrate-near cohesive failure (SCF) pattern. At higher laser powers the values decrease again and adhesive failure patterns show. The values of the contaminated surfaces follow this profile with a maximum at 30 % laser power and surpass the value of the uncontaminated and untreated surface with almost 5 MPa. Despite the high strength of the optimally treated uncontaminated sample not being reached, the failure patterns show substrate-near cohesive failure fractions (AF/SCF). Triple transit seems to be more effective and consistently yields

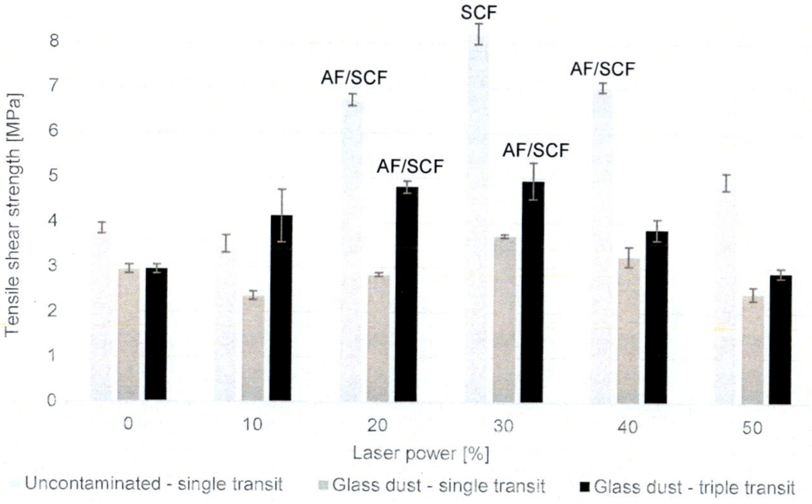

Fig. 7 Tensile shear strength with failure patterns (AF = adhesive failure, SCF = substrate-near cohesive failure)

higher tensile shear strength than single transit, which only lead to adhesive failure patterns.

4 Conclusion

The authors showed that a contamination that does not absorb laser radiation can still be cleaned off a delicate polymer surface in a laser process, without reducing functionality of the surface in regard to adhesive bonding. However, the contamination cannot be fully removed from the surface. After a certain threshold, in this study at a pulse fluence of 0.88 J/cm^2, the number of particles will not further decrease with higher laser power.

It was shown, that laser pre-treatment can significantly improve bonding strength at low pulse fluences. Higher pulse fluences damage the substrate and can reduce the bonding strength.

The laser cleaning process can substantially improve the bonding strength. Depending on the adhesive, the bonding strength can be improved to values above the initial, uncontaminated condition. Therefore, a laser cleaning process is a valid pre-treatment process for contaminated FRP surfaces. However, a complete cleaning is impossible, residues of the contamination have to be tolerated. Therefore, contaminated surfaces will not reach the same bonding strengths as uncontaminated and pre-treated surfaces. The pulse fluence for a cleaning process has to be chosen carefully, as high pulse fluences can damage the substrate and decrease adhesion properties of the adherend.

Acknowledgements The authors disclosed receipt of the following financial support for the research, authorship, and/or publication of this article: The IGF-project 19727, "Prozessbeobachtung und regelung der Klebvorbereitung PUR- und thermoplastbasierter, faserverstärkter Kunststoffe mittels Laser—ProKleb" of the research association "DVS German Welding Society" was funded within the framework of the industrial collective research program (IGF) by the Federal Ministry for Economic Affairs and Energy on the basis of a decision by the German Bundestag.

References

Dittmar, H., Weiland J., Wippo, V., Schiebahn, A., Jaeschke, P., Kaierl, S., Reisgen, U., Overmeyer, L.: Individualized and controlled laser beam pretreatment process for adhesive bonding of fiber-reinforced plastics—part II: automatic laser process control by spectrometry. J. Laser Appl. **33**, 012004 (2021).https://doi.org/10.2351/7.0000301

Fourrier, T., Schrems, G., Mühlberger, T., Heitz, J., Arnold, N., Bäuerle, D., Mosbacher, M., Bonebergs, J., Leiderer, P.: Laser cleaning of polymer surfaces. Appl. Phys. A **72**, 1–6 (2001). https://doi.org/10.1007/s003390000683

Kaelble, D.H.: Dispersion-polar surface tension properties of organic solids. J. Adhes. **2**(2), 66–81 (1970). https://doi.org/10.1080/0021846708544582

Kreling, S., Fischer, F., Delmdahl, R., Gäbler, F., Dilger, K.: Analytical characterization of CFRP laser treated by excimer laser radiation. Phys. Proc. **41**, 282–290 (2013). https://doi.org/10.1016/J.PHPRO.2013.03.080

Mittal, K.L., Lei, W.: Laser Technology: Applications in Adhesion and Related Areas. Scrivener Publishing, Beverly (2018)

Möller, D.: Luft – Chemie Physik Biologie Reinhaltung Recht. De Gruyter, Berlin (2003)

Napadlek, W.: Ablative laser cleaning of materials. J. KONES Powertrain Transp. **16**(1), 357–366 (2009)

Rabel, W.: Einige Aspekte der Benetzungstheorie und ihre Anwendung auf die Untersuchung und Veränderung der Oberflächeneigenschaften von Polymeren. Farbe Und Lack **77**(10), 997–1005 (1971)

Rauh, B., Kreling, S., Kolb, M., Geistbeck, M., Boujenfa, S., Suess, M., Dilger, K.: UV-laser cleaning and surface characterization of an aerospace carbon fibre reinforced polymer. J. Adhes. Adhesives. **82**, 50–59 (2018). https://doi.org/10.1016/j.ijadhadh.2017.12.016

Weiland, J., Dittmar, H., Beier, C.J.A., Ramesh, C., Marx, B., Schiebahn, A., Jäschke, P., Overmeyer, L., Reisgen, U.: Improvement of the adhesive bonding properties of a polyamide 6 injection molded fiber reinforced plastic component by laser beam pre-treatment. Proc. M. Eng. Part D: J. Automobile Eng. (2020). 10.1177%2F0954407020976195

3M Deutschland GmbH: 3M Glass Bubbles – Rundum Vielseitig – Merkmale, Spezifikationen und Anwendungen. https://docplayer.org/59007324-3m-tm-glass-bubbles-rundum-vielseitig-merkmale-spezifikationen-und-anwendungen.html (2017). Accessed on 06 May 2021

Structural Bonding of Thin-Walled Magnesium Die Cast Components

C. Koch, H. Sharabi, M. Kahlmeyer, A. Fros, S. Boehm, and M. Fehlbier

Abstract The demand for lighter, more fuel efficient and environment friendly vehicles has made the automotive manufacturer and suppliers to apply new technology and advanced materials. In this context, renewed interest arises in innovative applications of magnesium materials and requires reliable joining processes. Process-related contaminations, the high reactivity of magnesium alloys and their susceptibility to corrosion are major challenges for adhesive bonding. Therefore, structural bonding without pretreatment leads to low strength, high creep corrosion, and adhesive failure. Accordingly, the surface has to be pretreated to remove the instable magnesium oxide layer and release agent residues from previous processes. The main objective of this study is to investigate the suitability of laser cleaning and plasma polymerization as pretreatments for structural bonding of die cast magnesium alloys. The mechanical tests reveal a high lap shear strength for the laser pretreated samples. Additional deposition of adhesion-promoting layers by plasma could increase the aging resistance.

Keywords Adhesive–bonded joints · Magnesium · Pretreatment

C. Koch (✉) · M. Kahlmeyer · S. Boehm
Chair for Cutting and Joining Technologies (Tff), University of Kassel, Kassel, Germany
e-mail: christin.koch@uni-kassel.de

M. Kahlmeyer
e-mail: m.kahlmeyer@uni-kassel.de

S. Boehm
e-mail: s.boehm@uni-kassel.de

H. Sharabi · A. Fros · M. Fehlbier
Foundry Technology (GTK), University of Kassel, Kassel, Germany
e-mail: h.sharabi@uni-kassel.de

A. Fros
e-mail: adam.fros@uni-kassel.de

M. Fehlbier
e-mail: fehlbier@uni-kassel.de

L. F. M. da Silva and R. D. Adams (eds.), *6th International Conference on Adhesive Bonding 2021*, Proceedings in Engineering Mechanics,
https://doi.org/10.1007/978-3-030-87668-5_8

1 Introduction

Magnesium is the lightest of all structural applicable metals and has an excellent stiffness and strength-to-weight ratio (Campbell 2012). The high potential to produce thin-walled parts by die casting leads to versatile applications in many industries such as communication, automotive and aerospace (Czerwinski 2011). Magnesium is the least noble element in the galvanic series of the so called structural metals, making it highly susceptible to galvanic corrosion when it is in contact with almost all other metals. Moreover, magnesium forms a porous oxide layer under atmospheric humidity which offers insufficient protection against corrosion (Beck 2001). The protective ability of an oxide formed on a metal can be indicated by the Pilling–Bedworth ratio (PBR). The oxide layer is expected to be not protective if the PBR is less than 1.0. The PBR of Aluminium is greater than 1, which explains the protective effect of the aluminium oxide. Because the Pilling-Bedworth ratio for Magnesium is 0.81, the magnesium oxide film on the surface could not act as a protective layer. Therefore, magnesium alloys are mainly used for vehicle interiors and not for exterior parts.

In order to expand the areas of application, corrosion-resistant joining techniques play an essential role but are influenced by various factors. For example residues from the die casting process may affect the bonding strength of magnesium. Commonly used release agents for magnesium die casting are waxes, polysiloxanes and oils (Franke 2019). Furthermore, the weak adhering oxide layer of untreated magnesium (e.g. after die casting) forms a weak interphase which impairs the structural integrity of the joint. As a consequence, a surface treatment must be applied to create an appropriate surface for adhesive bonding and to increase corrosion resistance (Dorn and Anagreh 2007). So far, different coating technologies have been investigated for magnesium alloys including electrochemical plating, gas-phase deposition, conversion coating, anodizing and pickling (Gray and Luan 2002).

Chemical pretreatment methods are of the greatest importance for bonding of magnesium up to now. Chrome based pretreatments have shown excellent corrosion performance and provide a surface highly suitable for adhesive bonding (Forest Service U. S 1958). Liu et al. (2012) indicated that phosphate-permanganate surface treatment improves the static strength and thermal durability of the adhesively bonded AZ31 alloy. However, due to the toxicological reasons and long process sequences it is of utmost importance to develop alternative pretreatment methods.

Some studies have been carried out on the performance of mechanical pretreatment methods. Zheng et al. (2016) determined how surface polishing with silicon carbid sandpapers affects the surface characteristics of the AZ31B alloy and the adhesive strength. The results show that surface polishing decreases the surface free energy whereas the strength of the adhesive-bonded magnesium alloy is not significantly affected.

A laser pretreatment can enhance the adhesion on aluminium alloys. The reason stated is the enlarged surface caused by microroughness and the generated well-adhering oxide layer (Rechner et al. 2010a, b; Stammen et al. 2007). The effectiveness of plasma treatments on the adhesive bonding of aluminium was studied in Sperandio et al. (2010), Iorio et al. (1997), Dartevelle et al. (2003).

In contrast, the suitability of laser and plasma cleaning or plasma polymerization as pretreatment method for the bonding of magnesium alloys has been studied rarely yet. Schneider et al. (Schneider and Wrobel 2015) reported that a pretreatment of AZ31 alloy with pulsed high energy laser radiation leads to high peel strength (4.5 N/mm). Lap-shear tests and accelerated corrosion trials were not part of the study.

The intention of this paper is to describe the effects of laser cleaning and plasma treatments on the mechanical properties of adhesive bonded joints and to identify how release agents and the die casting process (e.g. form filling) affect the adhesive bond.

2 Experimental Details

For this work the magnesium alloy AZ91 was produced by hot chamber die casting, using an 580t device.

The following casting parameters were used:

Melt temperature (°C)	630
Piston speed (m/s)	1.3
Spray pressure release agent (bar)	1.5
Form temperature (°C)	220

To manufacture the required samples and to investigate the influence of the form filling dynamics on the bonding ability, a particular casting tool was manufactured (Fig. 1). The design results in different dynamics in each section. The outer parts are filled under laminar flow and should represent the *best case* conditions with high homogeneity and reduced impurity. The middle section is filled turbulently. For these *worst case* specimen more release agent residues are expected to remain on the magnesium surface. Furthermore it is estimated that the turbulent flow may result in higher porosity caused by entrapped gas.

To investigate the effect of the applied release agent on the bonding performance, three commercially available products with different chemical compositions were selected. RA1 is a water-based release agent cointaining a small amount of wax and oil, RA2 is water-based and contains a medium amount of waxes and oils, RA3 consists of wax, polysiloxane and oil. The release agents were sprayed in the mould in a concentration of 1:80 immediately before the casting process.

To remove the porous magnesium oxide layer and release agent residues a pretreatment of the magnesium surface was conducted by blasting or laser. For sandblasting corundum (particle size 110 μm) was accelerated by a pressure of 3.5 bar. The laser pretreatment was carried out by a pulsed Nd:YAG laser (wavelength 1064 nm) with a

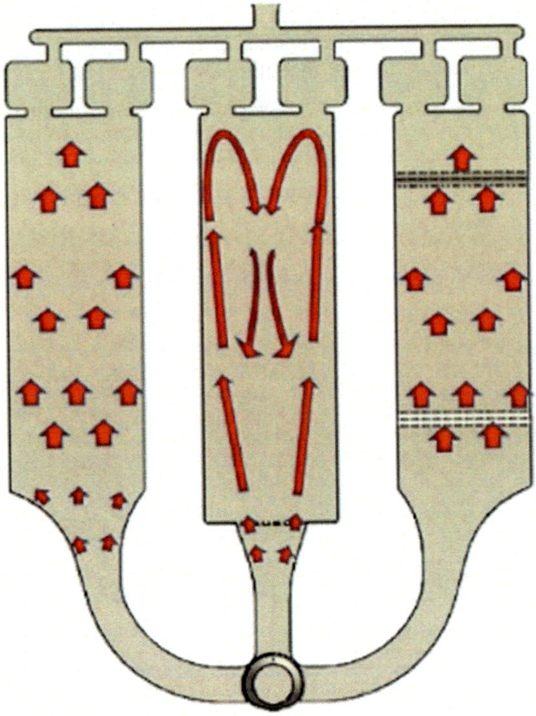

Fig. 1 Design of the mould. The outer sections are cast laminar, the middle part is cast turbulently

maximum mean laser power of 50 W and a pulse frequency of 50 kHz. For corrosion protection a hexamethyldisiloxane (HMDSO) layer was deposited by atmospheric plasma polymerisation after laser cleaning.

The bondings were performed with a room-temperature curing 2C epoxy resin (2C-EP) or a hot curing 1C epoxy resin (1C-EP).

The adhesive behaviour was studied by carrying out lap shear tensile tests according to DIN EN 1465. The sample geometry for the tensile shear test was $25 \times 100 \times 1.6\ mm^3$ with a reduced overlap length of 8 mm and an adhesive layer thickness of 0.3 mm. All tests were carried out at room temperature with a cross-head velocity of 10 mm/min using an universal testing machine.

The samples are tested in their unaged state or after ageing. For accelerated ageing the environmental cycle test PV1200 was carried out, where the joined specimen are exposed to a cyclic variing temperature profile between −40 and 80 °C over a period of 28 days.

3 Results and Discussion

3.1 *Influence of the Pretreatment*

After the die cast process release agents remain on the magnesium surface (casting skin) in different concentrations, which is very critical for adhesive bonding. Therefore, a pretreatment of the magnesium surface is indispensable, which reliably removes the casting skin as well as the natural oxide layer. Before bonding, the magnesium samples were subjected to the listed pretreatment methods, which remove surface contamination and/or have an activating and preserving effect on the adhesive bonding forces. To identify a suitable pretreatment method, the magnesium surface was analysed in the initial state and after the treatment. The characterisation was done by SEM/EDX and contact angle measurement. Figure 2 shows the SEM image (SE) and the EDX analysis of a laminar cast magnesium sample. One part of the sample is untreated and the other pretreated with a laser. It is obvious that the untreated part (with casting skin) has an optically inhomogeneous surface. Homogenization is achieved through laser cleaning.

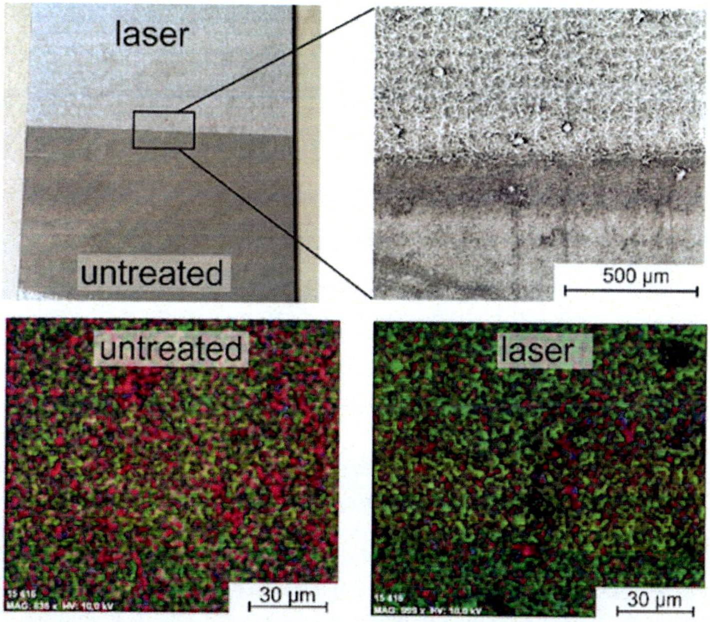

Fig. 2 SEM image (on top) and corresponding EDX-analysis (bottom) of an magnesium sample at the untreated and laser pretreated zone (green: magnesium, pink: carbon, purple: silicon)

Table 1 Contact angles and surface energies after different pretreatment methods

Pretreatment	Contact angle (H_2O) [°]	Surface energy [mN/m]
None	103.6 ± 7.7	31.5 ± 2.3
Sandblasting	39,1 ± 4.7	62,2 ± 2.3
Laser cleaning	26.3 ± 4.6	73.9 ± 2.2
Laser + plasma cleaning	–	–
Sandblasting + plasmapolym	82.6 ± 4.8	38.0 ± 4.1
Laser + plasmapolym	80.5 ± 4.9	36.0 ± 4.5

The laser induced a melting and evaporation of the magnesium surface. By recondensation of the evaporated material the surface is enlarged, furthermore the near-surface contaminations and release agent residues are removed.

To establish the effectivity of the laser pretreatment, it is of special interest if there are any release agent residues remaining on the surface. The elements carbon (Fig. 2, pink) and silicon (Fig. 2, purple) could be used as indicator because the release agent used consists of wax and polysiloxanes. The results show that the untreated parts are rich in carbon and silicon, in contrast in the laser pretreated zones the carbon and silicon content could be reduced.

The wettability was determined by contact angle measurements. The results of the laminar cast magnesium samples pretreated with different methods are shown in Table 1.

The initial magnesium sample has a contact angle of 104° in average, but it varies from 83° to 122° on the same sample which shows the inhomogeneity of the surface On the laser pretreated surface, the contact angle is reduced to 26°, thus the wettability is improved. This decrease is attributed to an enlargement of the surface area. If additional plasma cleaning is performed after laser treatment, the water droplet spreads immediately, so that no contact angle can be determined. This indicates very good wettability. The magnesium samples covered with plasma-polymerized hexamethyldisiloxane (PP-HMDSO) films have a contact angle around 80°.

Figure 3 illustrates the lap shear strength of specimens joined with 2C-EP and 1C-EP without and after pretreated with different methods.

The treatment of the magnesium surface by sandblasting (S), sandblasting with plasma-polymerization (S + L), laser (L) and laser and plasma polymerization (L + P) results in high lap shear strengths near the tensile strength of the specimen, while the untreated samples show a low lap shear strength around 10 MPa and a high standard deviation.

Despite a low surface energy of 36 mN/m the magnesium samples coated with a HMDSO-film by plasmapolymerization reach high lap shear strengths >22 MPa.

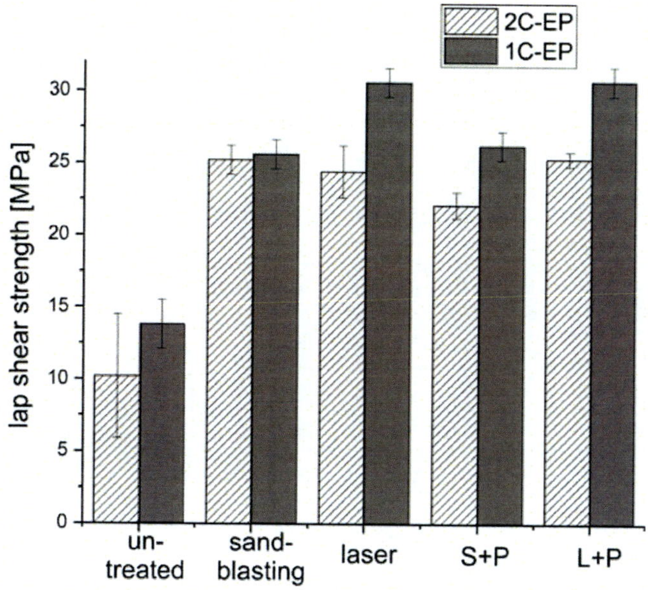

Fig. 3 Lap shear strength for magnesium samples pretreated with different methods and cast with RA1

This could be due to intermolecular interactions between the silicated oxid layer and the epoxy adhesive.

The failure modes of the specimens were evaluated by optical microscopy. It was found that the failure of the adhesive composites joined with 2C-EP was in the interphase between casting skin and the adhesive in case of the untreated samples. For the pretreated samples the amount of cohesive failure within the adhesive increased.

In the case of the adhesive bonds produced with 1C-EP, it can be seen that the lap shear strengths of the untreated specimens are higher compared with the cold-curing 2C-EP. This may result from the higher contamination tolerance of hot-curing 1 K epoxy resins. Nevertheless, a pretreatment enhances the lap shear strength.

3.2 Influence of the Release Agent

To analyse the influence of the release agent on the bonding strength, magnesium samples were cast with various release agents with different chemical composition and joined with 2C-EP.

If the magnesium surface is not pretreated, the lap shear strength of the samples, which were cast with RA3, is significant lower than the lap shear strength of the samples, which are cast with the other release agents (Fig. 4).

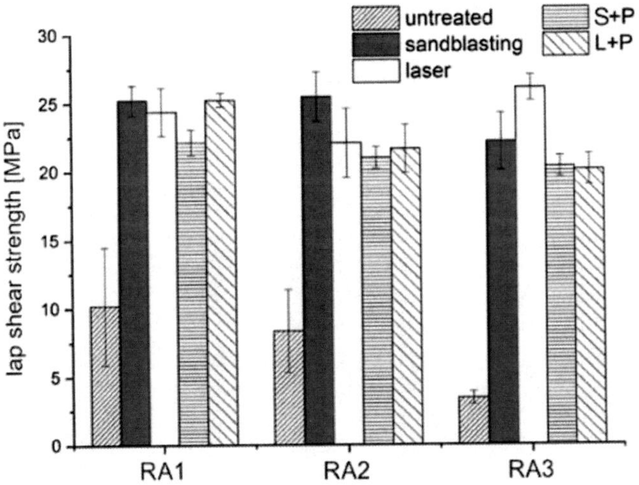

Fig. 4 Lap shear strength for magnesium samples cast with different release agents and joined with 2C-EP

RA3 contains a high amount of wax, polysiloxane and oil components, which may affect physico-chemical interactions and the adhesion. If the magnesium samples are pretreated by laser or sandblasted, similar lap shear strengths result, although different release agents are used. This shows that laser cleaning removes most of the release agent residues on the die casting surface. Thus, the adhesive suitability of samples containing a high proportion of organic compounds on the surface can also be increased.

3.3 *Influence of the Type of Flow*

In the case of industrial produced complex die-cast parts, some areas may be filled with a turbulent and others with a laminar flow. Therefore, the influence of the type of flow on the adhesive bonding was analysed. The laminar (*best case*) and turbulent (*worst case*) die cast samples were joined with 2C-EP.

As shown in Fig. 5, the worst case specimens exhibit a significantly lower lap shear strength than the best case specimens in the untreated state. After pretreatment, no significant differences can be observed between the worst case and best case specimens.

Turbulent flow lead to higher and inhomogeneous release agent distribution. However, the selected pretreatment methods are able to remove release agent residues with the casting skin, so that high lap shear strengths can be achieved independently from the mould filling.

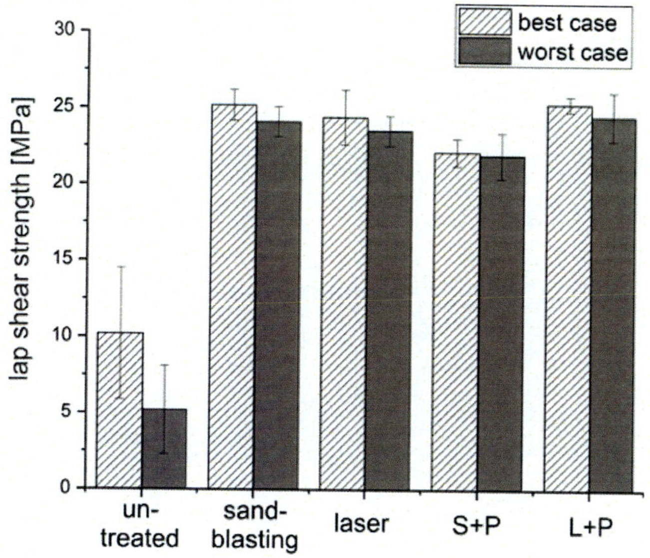

Fig. 5 Lap shear strength for magnesium samples cast with different mould filling. The samples were cast with RA1 and joined with 2C-EP

3.4 Environmental Testing

Environmental performance of the bonded joints was assessed by means of the environmental cycle test PV1200.

The initial fracture was cohesive for all pretreated magnesium samples. After aging, analysis of the fracture pattern presented adhesive failures for the untreated and sandblasted samples and cohesive failure for the laser treated samples. When the samples are additionally plasma polymerized, a predominantly failure in the magnesium oxide layer occur, as shown in Fig. 6.

After the environmental cycle test the untreated samples show a 50% lower lap shear strength (Fig. 7). While the sandblasted and laser treated samples display an approximately 20% lower lap shear strength than the unaged samples, an additional PP-HMDSO film led to the maximum lap shear strength after ageing.

4 Conclusions

The investigations described above show that laser pretreatment has a good cleaning effect on magnesium die cast samples. By removing the casting skin with release agent residues, it was possible to create a surface suitable for adhesive bonding.

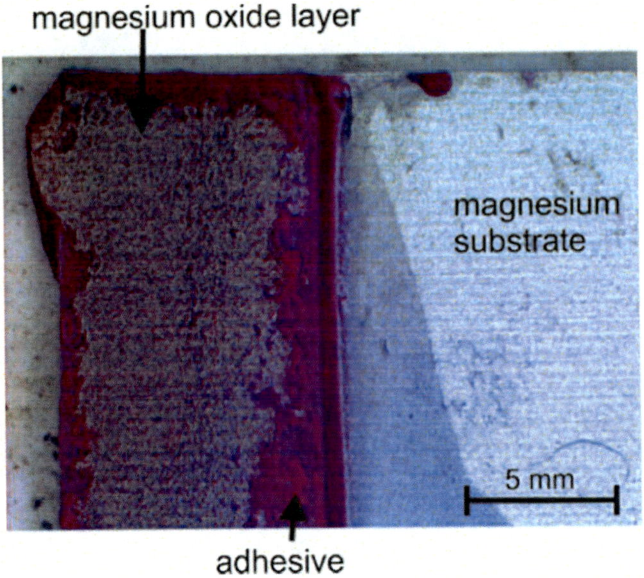

Fig. 6 Fracture pattern of the laser treated magnesium sample joined with 1C-EP

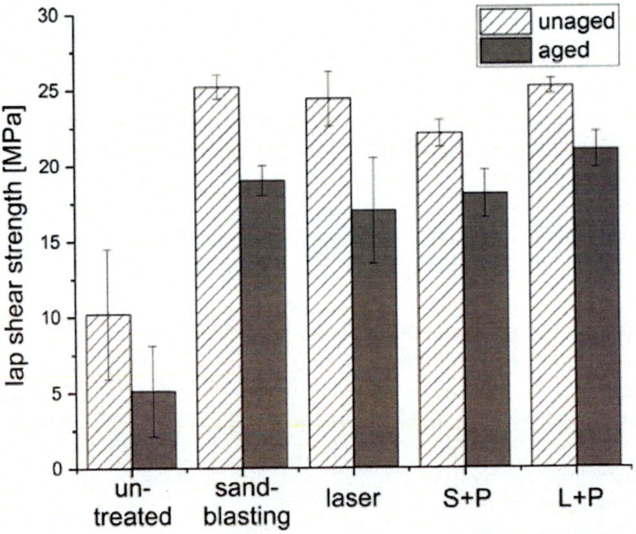

Fig. 7 Lap shear strength for magnesium samples after aging test. The samples were cast with RA1 and joined with 2C-EP

Depending on the release agent a reduced lap shear strength was detected. If magnesium die cast samples should be bonded, special attention should be paid to the components of the release agent. The best adhesive properties were found when the magnesium samples cast with a water-based release agent which contains only a small amount of wax and oil.

It has been found that the turbulent die cast samples present low adhesive strength in comparison to the laminar cast samples. However, the lap shear strength can be increased by appropriate pretreatment.

The results indicate that laser treatment and plasma polymerization are suitable for generating aging stable magnesium joints. Nevertheless, special focus should be placed on stabilizing the magnesium oxide layer in the future.

Acknowledgements The authors would like to thank the accompanying committee for their excellent support. The research project (AiF-No. 19873 N) of the DVS e.V. was funded through the program for the promotion of the "Industrial Community Research (IGF)" by the Federal Ministry of Economic Affairs and Energy through the AiF. This assistance is gratefully acknowledged.

References

Beck, A.: Magnesium und seine Legierungen. Springer Verlag, Berlin (2001)

Campbell, F.C. (ed.): Lightweight materials. Understanding the basics. ASM International, Ohio (2012)

Czerwinski, F. (ed.): Magnesium Alloys Corrosion and Surface Treatments. IntechOpen, Rijeka (2011)

Dartevelle, C., McAlpine, E., Thompson, G.E., Alexander, M.R.: Low pressure plasma treatment for improving the strength and durability of adhesively bonded aluminium joints. Surf. Coat. Technol. **173**(2–3), 249–258 (2003)

De Iorio, I., Leone, C., Nele, L., Tagliaferri, V.: Plasma treatments of polymeric materials and Al alloy for adhesive bonding. J. Mater. Process. Technol. **68**(2), 179–183 (1997)

Dorn, L., Anagreh, N.: Alterungsverhalten von Magnesium-Klebverbindungen nach unterschiedlicher Oberflächenvorbehandlung. Paper presented at the Große Schweißtechnische Tagung, Basel, 16–18 September 2007

Forest Service U. S. Departement of Agriculture, *Effect of Surface Treatment on the Adhesive Bonding Properties of Magnesium* (Forest Products Laboratory, Madison, 1958).

Franke, S.: Giesserei-Lexikon. Schiele & Schön GmbH, Berlin (2019)

Gray, J.E., Luan, B.: Protective coatings on magnesium and its alloys—a critical review. J. Alloys Compd. **336**, 88–113 (2002)

Liu, Z.-X., Sun, R.-N., Mao, Z.-P., Wang, P.-C.: Effects of phosphate pretreatment and hot-humid environmental exposure on static strength of adhesive-bonded magnesium AZ31 sheets. Surf. Coat. Technol. **206**(16), 3517–3525 (2012)

Rechner, R., Jansen, I., Beyer, E.: Influence on the strength and aging resistance of aluminium joints by laser pre-treatment and surface modification. Int. J. Adhes. Adhes. **30**, 595–601 (2010a)

Rechner, R., Jansen, I., Beyer, E.: Laser- und Plasmaverfahren im wirtschaftlichen Vergleich. Adhaes. Kleb. Dicht. **54**(1–2), 36–43 (2010b)

Schneider, N., Wrobel, C.: Use of high-energy laser radiation for surface preparation of magnesium for adhesive applications. Appl. Adhes. Sci. **3**(17), 1–9 (2015)

Sperandio, C., Bardon, J., Laachachi, A., Aubriet, H., Ruch, D.: Influence of plasma surface treatment on bond strength behaviour of an adhesively bonded aluminium-epoxy system. Int. J. Adhes. Adhes. **30**(8), 720–728 (2010)

Stammen, E., Dilger, K., Böhm, S., Hose, R.: Surface modification with laser: pretreatment of aluminium alloys for adhesive bonding. Plasma Process. Polym. **4**, 39–43 (2007)

Zheng, R., Lin, J., Wang, P.-C., Wu, Y.: Correlation between surface characteristics and static strength of adhesive-bonded magnesium AZ31B. Int. J. Adv. Manuf. Technol. **84**, 1661–1670 (2016)

The Behaviour of CFRP Adhesive Joints Under Different Loading Rates

Joana Filipa Dias Varajão, Ricardo João Camilo Carbas, Eduardo André Sousa Marques, and Lucas Filipe Martins da Silva

Abstract In recent years there has been an increasing interest in applying adhesive bonding to structural components of the vehicles manufactured by the automotive industry. When adhesive joints are used in automotive applications, additional factors such as the strength under impact loads play a decisive role. Under impact conditions, the bonded joints must provide sufficient strength to transmit the transient load without fracturing, ensuring the safety of the vehicle occupants. The main objective of this study was to evaluate the impact strength of single lap joints with carbon fibre reinforced plastic adherends, bonded using an epoxy adhesive. The joints were tested in a drop weight impact machine, in order to simulate impact conditions typical of the automotive industry. Quasi static-tests were also performed, allowing to compare the static and impact performance of the bonded joints. Numerical models, using cohesive zone elements were developed in ABAQUS® and the numerical results were validated with experimental results. Overall, the results of this study show that different overlap lengths considerably affect the impact response of the composite Single Lap Joints (SLJs). The results obtained with the use of analytical and numerical approaches were found to be in good agreement with the experimental data.

J. F. D. Varajão · L. F. M. da Silva
Departamento de Engenharia Mecânica, Faculdade de Engenharia, Universidade Do Porto (FEUP), Rua Dr. Roberto Frias, 4200-465 Porto, Portugal
e-mail: em10142@fe.up.pt

L. F. M. da Silva
e-mail: lucas@fe.up.pt

R. J. C. Carbas (✉) · E. A. S. Marques
Institute of Science and Innovation in Mechanical and Industrial Engineering (INEGI), Rua Dr. Roberto Frias, 400, 4200-465 Porto, Portugal
e-mail: rcarbas@fe.up.pt

E. A. S. Marques
e-mail: emarques@fe.up.pt

L. F. M. da Silva and R. D. Adams (eds.), *6th International Conference on Adhesive Bonding 2021*, Proceedings in Engineering Mechanics,
https://doi.org/10.1007/978-3-030-87668-5_9

1 Introduction

Adhesively bonded joints are increasingly being used in the automotive industry. However, these applications require the bonded parts to exhibit a good resistance to impact loads, allowing to ensure the safety of the occupants of a vehicle during a collision. Therefore, both the adhesives and adhesive joints must be able to provide high impact resistance and deform significantly without breaking. By keeping the components joined, the bonded joints ensure that the adherends can also deform during the collision, absorbing large amounts of energy (Adams 1984).

Furthermore, the need to reduce the vehicles weight is also growing, leading to an increase in the use of lightweight materials and construction technologies, such as composite materials. Composite materials are combinations of two or more different materials and are characterized by their constituent mechanical properties, generally exhibiting excellent specific stiffness and strength. When bonded, their performance is vastly different from that exhibited by metals, which has made them target of many research works. However, most of the these works focus on quasi-static performance, and very few studies can be found in literature on the behaviour of composites under impact loads (da Silva et al. 2018).

Within the subject of bonded composite joints, the automotive industry is especially concerned by the relatively low delamination resistance of Carbon Fibre Reinforced Plastics (CFRP) laminates and/or the fracture behaviour of joints. Thus, different impact tests have been generally used to quantify the maximum load and the energy absorbed in adhesive joints with composite adherends. Impact tests can be categorized into three ranges: low velocity, medium velocity and high velocity. In a laboratory setting, an impact can be simulated using essentially three tests: pendulum type, where tests rates are below 10 m/s, drop-weight tests, with rates up to 10 m/s and the split Hopkinson pressure bar, which can provide rates of up to 100 m/s.

In the last few decades, diverse works in the literature have identified the importance of studying the behaviour of CFRP at high strain rates in order to observe the damage and the mechanisms of failure encountered under impact conditions. Many studies (da Silva et al. 2018) concluded that the interlaminar shear strength of carbon fibre/epoxy composites increases with strain rate. Hosur et al. (2005) explained that this behaviour can be attributed to the time-dependent deformation of the material, which is a product of the visco-elastic behaviour of the matrix in the composite. Under impact conditions, it is necessary to ensure that the adhesive layer retains sufficient toughness to induce the damage mechanisms that might occur in composite or layered materials. The manner in which composite materials respond to impact loading is also very different than that of metals (Machado et al. 2019). For low and intermediate incident energies, metals absorb energy through elastic and plastic deformation (dos Reis et al. 2021). However, in the case of composites, the ability to undergo plastic deformation is extremely limited, and the absorbed energy creates large, fractured areas, with ensuing reductions in both strength and stiffness (Reis et al. 2020). Furthermore, while metal structures collapse under impact via buckling processes involving extensive plastic deformation, composites fail through a sequence of fracture mechanisms. The actual mechanisms and sequence of damage are highly dependent on the geometry of the structure, lamina orientation and loading speed, all of which can be suitably designed to develop high-energy absorbing mechanisms (Hiley et al. 1997).

The behaviour of adhesives under impact has also been studied. Harris and Adams (1985) characterized adhesive performance at high loading rates with a pendulum impact machine. The results showed that for a wide range of adhesives, joint strength is not significantly affected by high rates of loading. The second conclusion obtained was the existence of a relation between adherend strength, joint strength and energy absorption. They noticed that using a rubber-toughened adhesive with ductile aluminium alloy adherends provided the highest energy absorption, while for a high-strength aluminium alloy, the joint strength increased with a reduction of the energy absorption due to the high yield strength of the aluminium. When an adhesive is strong enough to keep the adherends together under impact load, it is also the adherend yielding that controls failure. Can be conclude that the energy absorption arises not from the adhesive, but from the deformation of the adherend material and that a strong joint does not necessarily have optimum energy absorbing capabilities (da Silva et al. 2012). When a ductile and strong adhesive is used, the sensitivity of the adherend material to high strain-rate is what determines the failure process (Avendaño et al. 2016a, b).

It is also known that when joints, such as SLJs, are submitted to high strain-rate conditions, their material properties tend to be different from the ones determined under quasi-static loading conditions. Many experimental results for impact tests can be found in the literature, but if the materials have the same properties as those obtained in static conditions, it becomes possible extrapolate the materials properties in order to predict failure (Zgoul and Crocombe 2004). It is also reported that adhesive joints bonded with an epoxy adhesive are stronger than those with a polyurethane adhesive and there is an increase in joint strength under both static and impact loads (Loureiro et al. 2010). When mild steel adherends are used and adherend yielding is significant, the properties of the steel are those who determine the failure of SLJs at room temperature (Viana et al. 2018). Tests on SLJs bonded with a high elongation epoxy adhesive and ductile adherends show that the failure mode obtained under high strain-rate was almost the same as the one obtained in quasi-static. However, some differences in the behaviour of joints were reported—the adherends deformed less than in quasi-static tests and absorbed less energy. Consequently, the failure load increased but the adhesive experienced similar damage as in the quasi-static tests (Saldanha et al. 2013).

This study has the main goal to study the response of CFRP adhesive joints under different rates, evaluating both experimentally and numerically the failure process and mechanical performance of the joint, ensuring its suitability for automotive industry applications.

2 Experimental Details

2.1 Adhesive

A two-component epoxy adhesive, SikaPower 4720 supplied by Sika® Portugal, SA (Porto, Portugal) was used in this work. The cure cycle used consisted of a stage at room temperature during one full day.

2.1.1 Tensile Tests

The test most commonly used to determine the strength properties of adhesives is the tensile test carried out on bulk specimens. The properties extracted from this test are intrinsic to the material and can be obtained under a uniform and uniaxial state of stress, with no influence of the adherends. Nevertheless, the properties determined may still vary from the in situ (in a joint) tests. The test geometry should provide a uniformly distributed stress across the contact surface as to minimize the stress concentrations. In general, the geometry used for the tensile specimens is a dogbone-shaped specimen that follows the EN ISO 527-2 standard (da Silva et al. 2012). The specimen geometry used in this work is shown in Fig. 1.

The bulk tensile tests were performed in an INSTRON® model 3367 universal test machine (Norwood, Massachusetts, USA) with a load cell capacity of 30 kN. The adhesive dogbone specimens were tested at room temperature, under a constant displacement rate of 1 and 100 min/min. Three tests were performed for each condition. Strain was determined using an optical method, with a digital camera monitoring the separation of the two lines inscribed on the test specimen.

Figure 2 shows the results of the tensile test for quasi-static testing and for 100 mm/min. An increase of adhesive tensile strength at 100 mm/min can be seen, as well as a minor variation in the average Young's modulus. The ductility at 100 mm/min was found to be lower than that obtained for 1 mm/min, which is consistent with the data reported in the literature (da Silva et al. 2018). The tensile specimen at 1 mm/min were performed in previous study (Costa et al. 2016) and

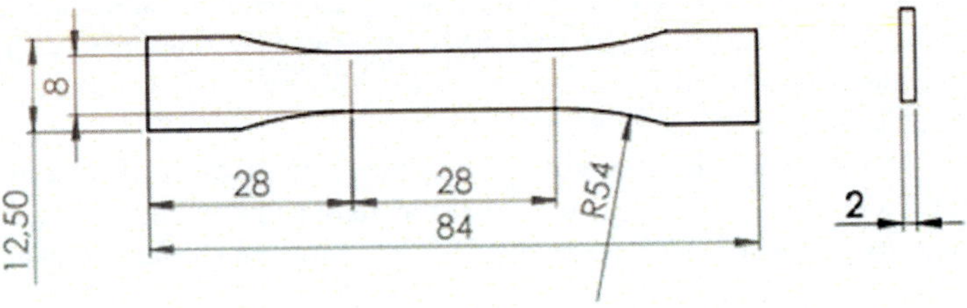

Fig. 1 Geometry of the bulk tensile test specimen (dimensions in mm)

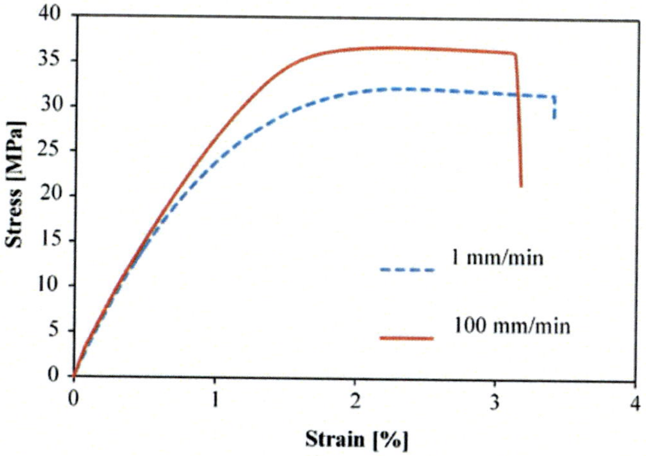

Fig. 2 Comparison stress–strain curves at different rates (1 and 100 mm/min)

for 100 mm/min a Young's modulus of 2431 ± 90.0 MPa was obtained, a tensile strength of 35.7 ± 1.7 MPa and strain to failure of 3.1 ± 0.4.

2.1.2 Thick Adherend Shear Test (TAST)

The TAST is one of the most popular types of failure strength test because it uses an easy to manufacture specimen and requires a relatively straightforward testing procedure (ASTM D3983). In this work, a bondline thickness of 0.5 mm was used and an overlap length of 5 mm, with the specimens being manufactured in a mould designed to control properly the final dimensions of specimens, as it can be seen in Fig. 3 (da Silva et al. 2012). The tests were performed at room temperature and

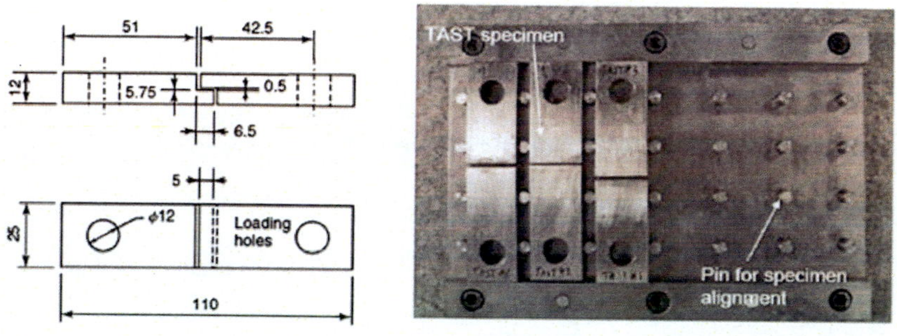

Fig. 3 Specimen geometry used for TAST tests and the mould used to manufacture the TAST specimens (da silva et al. 2012)

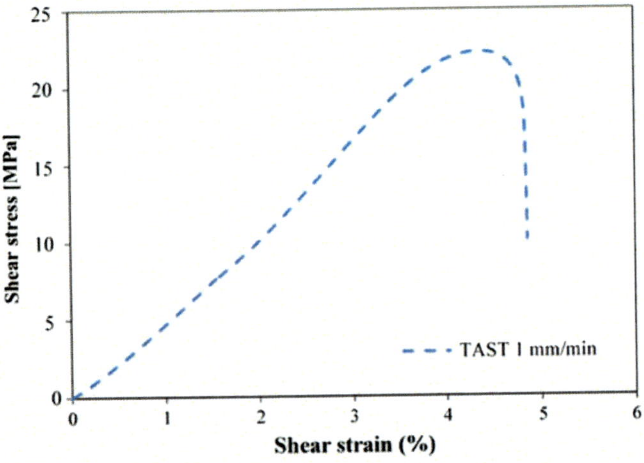

Fig. 4 Shear stress–strain curve of SikaPower 4720 at 1 mm/min

a constant crosshead rate of 1 mm/min. Tests at 100 mm/min were not possible, since the greater brittleness of the adhesive did not allow to measure the strain accurately. The displacement was measured with an extensometer, directly mounted in the metallic substrate. Three joints were tested for each adhesive.

Typical shear stress–strain curves of SikaPower 4720 tested at room temperature are shown in Fig. 4. From the shear stress–strain curves, the shear modulus and shear strength were calculated. The values of shear modulus were calculated from the tangent to the shear stress–strain curve at the origin. For 1 mm/min, a shear modulus of 804 ± 43.4 MPa was obtained, with a shear strength of 22.9 ± 0.7 MPa and a shear strain to failure of 5.2 ± 0.3.

2.2 *Adherends*

The material used to manufacture the adherents of the SLJs consisted of unidirectional CFRP laminates. The main mechanical properties of the unidirectional 0° lay-ups of carbon/epoxy pre-preg (SEAL® Texipreg HS 160 RM) used are presented in Table 1.

Table 1 Properties of the CFRP elastic orthotropic used for the SLJs (Campilho et al. 2005)

Elastic modulus (MPa)	Poisson's ratios	Shear modulus (MPa)
E_1= 1.09E5 E_2= 8819 E_3= 8819	$\nu_{12} = 0.342$ $\nu_{13} = 0.342$ $\nu_{23} = 0.380$	$G_{12} = 4315$ $G_{13} = 4315$ $G_{23} = 3200$

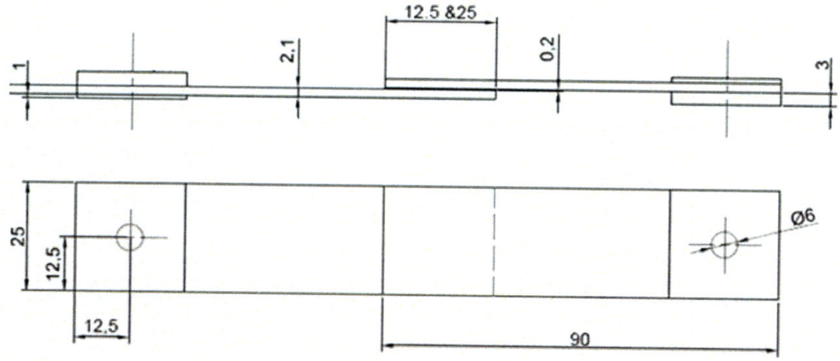

Fig. 5 SLJs geometry used in static and impact tests (dimensions in mm)

2.3 Joint Geometry

Two groups of SLJs were used, with a single difference between them lying in the overlap lengths, which were 12.5 and 25 mm. Another different aspect is the presence of aluminium and steel tabs, all of which were bonded in the ends of the adherends for the impact tests. This reinforcement is done to reduce the likelihood of failure in the attachment to the impact testing setup, where a special assembly is needed. The composite adherends thickness was 2.1 mm and the adhesive thickness used was 0.2 mm. The geometry used for static and impact tests was chosen because it is generally representative of structural adhesive joints found in the automotive industry. The specimen geometry can be seen in Fig. 5.

2.4 Test Conditions

The static tests were performed in the universal testing machine INSTRON® model 3367 (Norwood, Massachusetts, USA) with a capacity of 30 kN with a constant crosshead rate of 1 mm/min at room temperature. Four joints were tested for each condition and for each joint tested load–displacement curve was produced.

The SLJs tested under in the impact loads were performed in a Rosand® Instrumented Falling weight impact tester, type 5 H.V. (Stourbridge, West Midlands, U.K.). This machine drops a guided mass from a certain high until it impacts on the device that holds the specimen. The energy applied in the impact is controlled by the weight of the falling mass, and the speed can be set by the height of the drop. The load is transmitted to the SLJs thanks to a vertically guided anvil. A load cell, attached to the moving impactor, is used to measure load and time. After assembling the specimen, the mass (26 kg) and drop heigh were set to provide an impact energy of 50 J and an

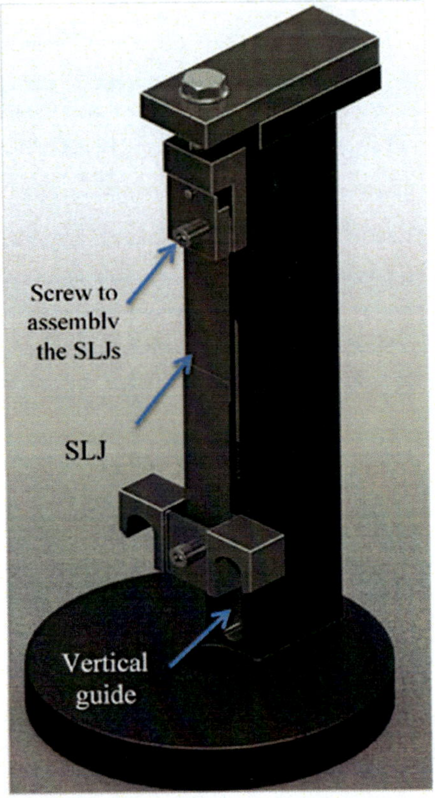

Fig. 6 Setup for impact testing of the joints under impact loads (Avendaño et al. 2016a)

impact speed of 2 m/s. Figure 6 shows the experimental setup used for SLJs impact tests. Five SLJs were tested for each adhesive type and SLJs overlap length.

3 Numerical Details

The ABAQUS® software was selected to perform the numerical analysis in order to predict the strength of SLJs. The mesh used for the single-lap joints can be seen in Fig. 7. The mesh was particularly refined at the overlap edges in order to accurately capture locations of stress concentration. To simplify the model, the joint was modelled as a two-dimensional construct, with plane-strain solid elements.

In this work, a continuum-based approach was considered to model the finite thickness of the adhesive layer (da Silva and Öchsner 2008). The adhesive layer was modelled using with Cohesive Zone Modelling (CZM), with a traction–separation law representing the adhesive properties. The adherends were reproduced as elastic

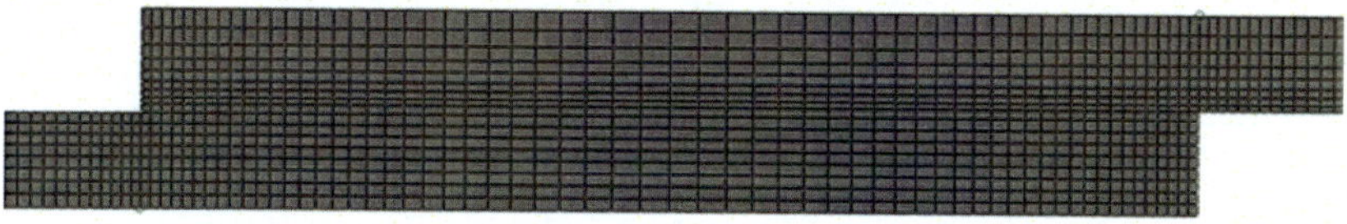

Fig. 7 Typical mesh refinement used in models

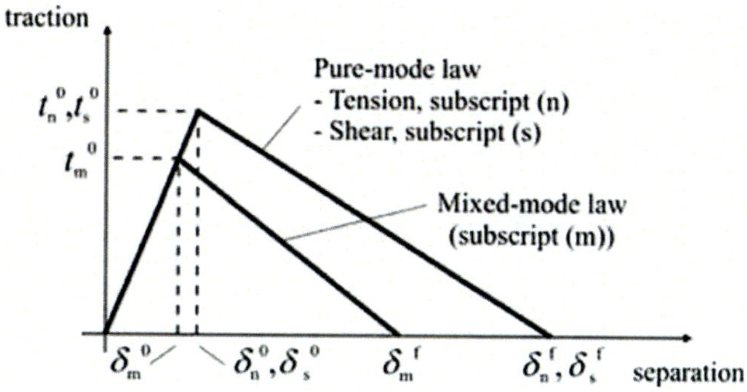

Fig. 8 Traction–separation law with linear softening law available in ABAQUS®

solids since they do not undergo damage. The triangular CZM (see Fig. 8) formulation was chosen for this analysis because of its simplicity, ample use in research and due to the fact that it is fully integrated in ABAQUS, allowing for a mixed-mode formulation.

3.1 *Static Analysis*

The location of the elements introduced in the Abaqus model are shown in Fig. 9. The joints were restrained (i.e., clamped) at one of the edges to simulate the clamping conditions in the machine grips while the other edge was subjected to a tensile

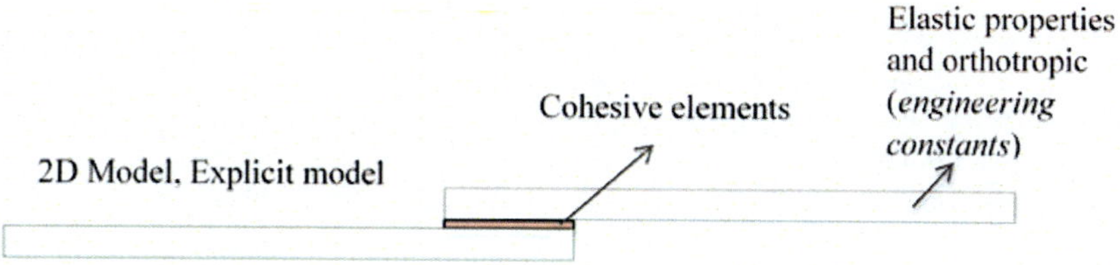

Fig. 9 Elements used in the static model

displacement with transverse restraining. The G_{IIC} of adhesive was obtained via an iterative estimation based on numerical and experimental data.

3.2 *Impact Analysis*

In order to predict and simulate the mechanical response of the composite joint under impact loads, an explicit model was used. An explicit model provides the capability to analyse high-speed, time dependent problems, such as drop weight tests and crash analyses of structural members. To simulate the energy imposed during testing in a drop weight tester, a mass was introduced in the extremity of joint, as shown in Fig. 10.

The boundary conditions were the same as the ones used in a static analysis. The only exception is the establishment of a predefined velocity field, with a value of 2 m/s. This value was introduced in order to obtain the same conditions found in drop weight tests.

Table 2 shows the properties used for the establishment of the adhesive layer damage laws. The properties of the adhesive at impact speed were estimated via an extrapolation process, based on the experimental properties obtained at 1 and 100 mm/min. Table 2 presents the values calculated with extrapolation for

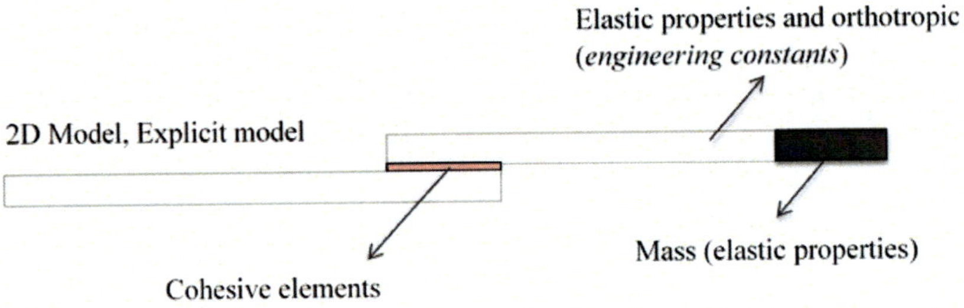

Fig. 10 Elements used in the impact model

Table 2 Mechanical properties of the adhesive SikaPower® 4720 used in numerical analysis (Costa et al. 2016)

Property	1 mm/min	2 m/s
Young's modulus (MPa)	2170	3113
Shear modulus (MPa)	804	1198
Normal stress (MPa)	32.0	45.4
Shear stress (MPa)	22.9	30
Critical strain energy release rate in pure mode I (N/mm)	1.63	1.63
Critical strain energy release rate in pure mode II (N/mm)	3.0	4.0

2 m/s (impact speed). A logarithmic extrapolation process (Propertytodetermine $=$ A$ln(\dot{\varepsilon})$ + B) was used in order to obtain the CZM parameters, where A and B are constants obtained to experimental data and $\dot{\varepsilon}$ is the strain rate in s^1.

4 Results and Discussion

4.1 Experimental Results—Static Tests

The load–displacement curves of the SikaPower 4720 for two overlap lengths (12.5 and 25 mm) are depicted in Fig. 11. The load and displacement data used were extracted from the recorded files provided by the universal testing machine. The results show that the joint strength increases with the overlap length, but the maximum load is doubled for the 25 mm overlap. Both overlap lengths under analysis exhibited cohesive failure in adhesive.

4.2 Experimental Results—Impact Tests

The results of the impact are presented in Fig. 12, showing the values of load and displacement acting on the joint. The load and time are obtained from the load cell, which registers the values during the test and the machine software then calculates the displacement by integration of the acceleration data. Load–displacement (P–δ)

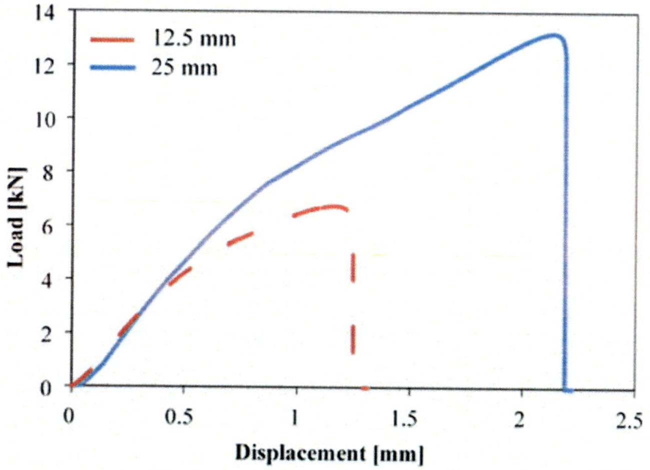

Fig. 11 Quasi-static response of SikaPower 4720 SLJs for two overlap lengths

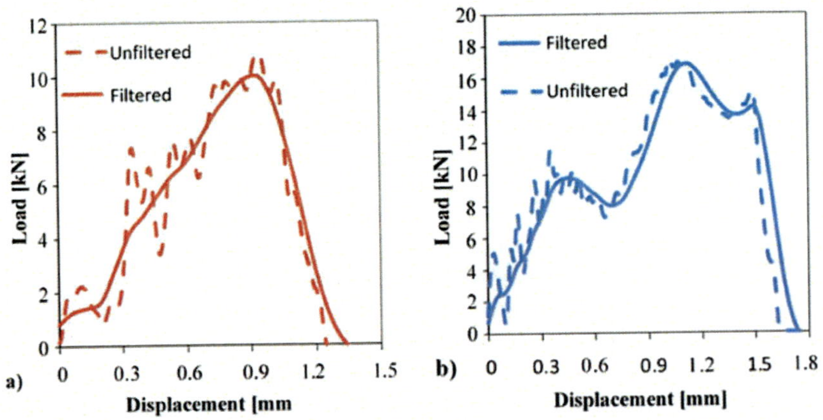

Fig. 12 Load versus displacement curves under impact loads, **a** 12.5 mm and **b** 25 mm

curves can then be obtained. A digital filter, with a value of 3 kHz (within the limits recommended by the machine supplier) was applied to all the curves in order to filter the oscillations provided by the impact machine (drop weight) and to allow for a better estimate of the effective load of each specimen.

Figure 12 shows the representative load variation of the impact tests using SLJs bonded with SikaPower 4720 adhesive. The displacement is equivalent to the total elongation of the specimens, obtained from the load vs. elongation curve. For all specimen's cohesive failure in the adhesive was obtained.

4.3 *Numerical Joint Strength Prediction*

This section describes the results of the numerical analysis carried out on the CFRP joints. Figure 13 gives an example of the overall correlation obtained between the experimental and numerical (CZM) $P\delta$ curves for quasi-static results and for an overlap length of 25 mm.

In order to select the most accurate value of Critical strain energy release rate in pure mode II (G_{IIC}) for use in the impact numerical simulations, an iterative approach was followed, adjusting the value of this property until a peak force closer to the experimental value was encountered. In the case of 25 mm overlap length, this corresponds to a load of 17.3 kN. An example that was done for each value of G_{IIc} is shown in Fig. 14.

Table 3 shows the values obtained for failure load for three fracture toughness in mode II used in the numerical model. The value found for G_{IIC} that is more appropriate was 5 N/mm.

Figure 15 gives an example of a comparison between the experimental and numerical curves for 25 mm of overlap length. A difference between the experimental and

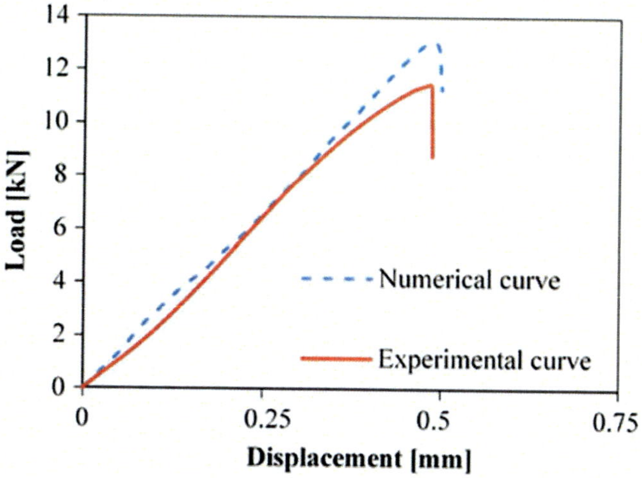

Fig. 13 Experimental versus numerical under static loads

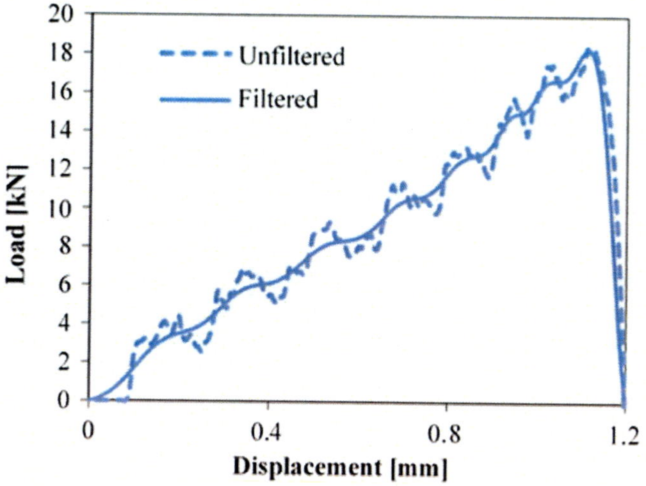

Fig. 14 Impact curve obtained in Abaqus simulation for $G_{IIC} = 5$ N/mm

Table 3 Results obtained in the numerical model for various values of G_{IIC}

G_{IIc} (N/mm)	P_{max} (kN)
5	18.2
4	17.1
3	16.8

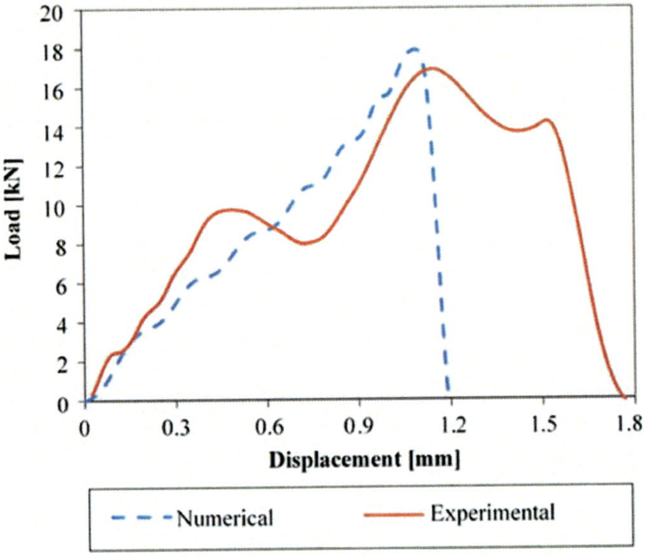

Fig. 15 Experimental versus numerical impact results of SLJs bonded with SikaPower® 4720

numerical displacements is evident. In the case of experimental results, the machine measures the displacements of all their components which explains a large difference between numerical and experimental values.

Figure 16 show a comparison between the experimental and numerical failure loads and the predictions obtained using analytical method for the joints. For the analytical joint strength prediction, the Adams et al. (1997) model was chosen. The global yielding criteria was found to provide reasonable values. However, when a numerical CZM model was used to predict the strength of the joints better results were obtained. The CZM model identified a cohesive failure mode for both overlap length studied as observed in the experiments.

5 Conclusions

In order to understand the behaviour of CFRP adhesive joints under impact conditions, aiming at an application in the automotive industry, a theoretical and experimental analysis is required, particularly in the field of the strain rate dependence of structural adhesives. In line with the objective of this work, static and impact SLJs tests were performed with the same adherends and using the adhesives SikaPower 4720. In order to validate the experimental results, numerical simulations of the CFRP adhesive joints were carried out. For the numerical modelling and simulation,

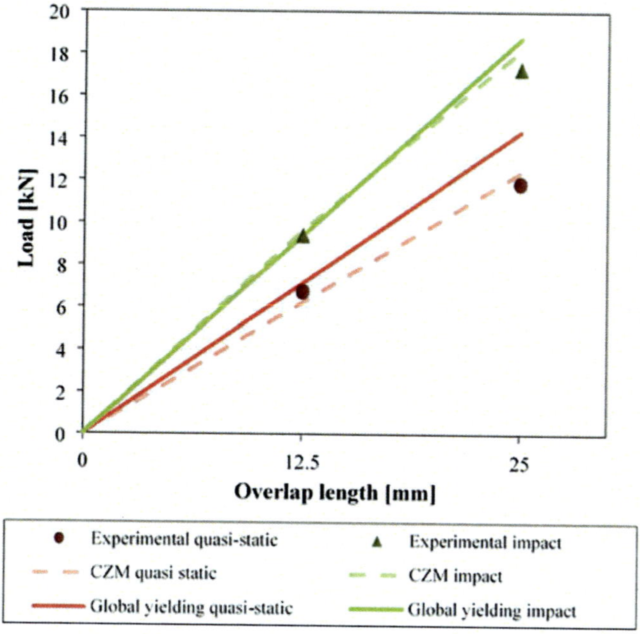

Fig. 16 Experimental results versus prediction of SLJs bonded with SikaPower® 4720

ABAQUS® software was used. The mesh creation, definition of boundary conditions and definition of material properties were part of the modelling process.

Therefore, in terms of impact tests it was concluded that the high strain rate considerably affects the responses of the SLJs when compared to static conditions. The effect of the overlap length was found to be proportional to the bonded area in both quasi-static and impact tests. The results obtained for numerical simulations agree with the experimental data and with the analytical predictions for SLJs with SikaPower 4720. The logarithmic extrapolation used in this case to obtain the adhesive properties under impact allowed to obtain a reasonably accurate approximation of the adhesive strength at high strain rates.

References

Adams, R.: Structural Adhesive Joints in Engineering. Springer, Netherlands, London (1984)

Adams, R.D., Comyn, J., Wake, W.C.: Structural Adhesives Joints in Engineering, 2nd edn. Chapamn & Hall, London (1997)

Avendaño, R., Carbas, R.J.C., Marques, E.A.S., da Silva, L.F.M., Fernandes, A.A.: Effect of temperature and strain rate on single lap joints with dissimilar lightweight adherends bonded with an acrylic adhesive. Compos. Struct. **152**, 34–44 (2016a). https://doi.org/10.1016/j.compstruct.2016.05.034

Avendaño, R., Carbas, R.J.C., Chaves, F.J.P., da Silva, L.F.M., Fernandes, A.A.: Impact loading of single lap joints of dissimilar lightweight adherends bonded with a crash-resistant epoxy adhesive. J. Eng. Mater. Technol. **138**(4), 041019 (2016b).https://doi.org/10.1115/1.4034204

Campilho, R.D.S.G., de Moura, M.F.S.F., Domingues, J.J.M.S.: Modelling single and double-lap repairs on composite materials. Compos. Sci. Technol. **65**(13), 1948–1958 (2005). https://doi.org/10.1016/j.compscitech.2005.04.007

Costa, M., Viana, G., Canto, C., da Silva, L.F.M., Banea, M., Chaves, F., Campilho, R., Fernandes, A.A.: Effect of the size reduction on the bulk tensile and double cantilever beam specimens used in cohesive zone models. Proc. IMechE Part L: J. Mater. Des. Appl. **230**(5), 968–982 (2016). https://doi.org/10.1177/1464420715610248

da Silva, L.F.M., Öchsner, A.: Modeling of Adhesively Bonded Joints. Springer, Berlin (2008)

da Silva, L.F.M., Dillard, D., Blackman, B., Adams, R.: Testing Adhesive Joints Best Practices. Wiley-VCH Verlag GmbH & Co. KGaA, Weinheim (2012)

da Silva, L.F.M., Öchsner, A., Adams, R.: Handbook of Adhesion Technology, 2nd edn. Springer International (2018)

dos Reis, M.Q., Carbas, R.J.C., Marques, E.A.S., da Silva, L.F.M.: Functionally graded adhesive joints under impact loads Proc. IMEchE Part D J. Automobile Eng. (2021).https://doi.org/10.1177/09544070211004505

Harris, J.A., Adams, R.D.: An assessment of the impact performance of bonded joints for use in high energy absorbing structures. Proc. Inst. Mech. Eng. Part C J. Mech. Eng. Sci. **199**(2), 121–131 (1985). https://doi.org/10.1243/PIME_PROC_1985_199_102_02

Hiley, M.J., Dong, L., Harding, J.: Effect of strain rate on the fracture process in interlaminar shear specimens of carbon fibre-reinforced laminates. Compos. Part A Appl. Sci. Manuf. **28**(2), 171–180 (1997). https://doi.org/10.1016/S1359-835X(96)00112-1

Hosur, M.V., Waliul Islam, S.M., Vaidya, U.K., Kumar, A., Dutta, P.K., Jeelani, S.: Dynamic punch shear characterization of plain weave graphite/epoxy composites at room and elevated temperatures. Compos. Struct. **70**(3), 295–307 (2005). https://doi.org/10.1016/j.compstruct.2004.08.041

Lifshitz, J.M., Leber, H.: Response of fiber-reinforced polymers to high strain-rate loading in interlaminar tension and combined tension/shear. Compos. Sci. Technol. **58**(6), 987–996 (1998). https://doi.org/10.1016/S0266-3538(97)00226-1

Loureiro, A.L., da Silva, L.F.M., Sato, C., Figueiredo, M.A.V.: Comparison of the mechanical behaviour between stiff and flexible adhesive joints for the automotive industry. J. Adhes. **86**(7), 765–787 (2010). https://doi.org/10.1080/00218464.2010.482440

Machado, J.J.M., Hayashi, A., Nunes, P.D.P., Marques, E.A.S., Carbas, R.J.C., Sato, C., da Silva, L.F.M.: Strain rate dependence of a crash resistant adhesive as a function of temperature for the automotive industry. Proc. IMechE Part L: J. Mater. Des. Appl. **233**(11):2189–2203 (2019). https://doi.org/10.1177/1464420719836914

Reis, J.M.L., Machado, J.J.M., Marques, E.A.S., Carbas, R.J.C., da Silva, L.F.M.: Displacement rate effect in the fracture toughness of glass fiber reinforced polyurethane. J. Compos. Mater. **54**(22), 3047–3054 (2020). https://doi.org/10.1177/0021998320908300

Saldanha, D.F.S., Canto, C., da Silva, L.F.M., Carbas, R.J.C., Chaves, F.J.P., Nomura, K., Ueda, T.: Mechanical characterization of a high elongation and high toughness epoxy adhesive. Int. J. Adhes. Adhes. **47**, 91–98 (2013). https://doi.org/10.1016/j.ijadhadh.2013.08.001

Viana, G., Machado, J., Carbas, R.J.C., Costa, M., da Silva, L.F.M., Vaz, M., Banea, M.D.: Strain rate dependence of adhesive joints for the automotive industry at low and high temperatures. J. Adhes. Sci. Technol. **32**(19), 2162–2179 (2018). https://doi.org/10.1080/01694243.2018.1464635

Zgoul, M., Crocombe, A.D.: Numerical modelling of lap joints bonded with a rate dependent adhesive. Int. J. Adhes. Adhes. **24**(4), 355–366 (2004). https://doi.org/10.1016/j.ijadhadh.2003.11.006

Long-Term Resistant Bonding Between Wood and Aluminium Alloy for Hybrid Lightweight Composites

M. Burnett, M. Kahlmeyer, N. Ratsch, and S. Böhm

Abstract Wood-based composites offer great lightweight potential for numerous applications and contribute to greater sustainability. Of particular interest are hybrid wood-metal composites, which combine stiffness and strength of metals with the low density of wood. This study investigates the durability of adhesive joints between wood and aluminium for advanced technical applications. Therefore, selected plywoods and aluminium substrates were pretreated and glued by a PUR and a water-based PVAc adhesive. The joints were tested under tensile and bending loads as a function of various climates. In addition to the mechanical properties, the influence of surface treatments on the adhesion of the adhesives was examined. The results demonstrate the necessity of pre-treatment and how the durability of hybrid wood-metal composites can be enhanced and motivate the use of sustainable adhesives for wood-based composites.

Keywords Adhesive bonded hybrid wood-metal joints · Sandwich composite · Surface pre-treatment

1 Introduction

Due to the outstanding density-related mechanical properties and ecological and economical benefits, wood offers extraordinary potential for lightweight construction and CO_2 savings. But until now it has not yet been exhausted for structural high

M. Burnett (✉) · M. Kahlmeyer · N. Ratsch · S. Böhm
Department for Cutting and Joining Technologies (tff), University of Kassel, Kassel, Germany
e-mail: m.burnett@uni-kassel.de

M. Kahlmeyer
e-mail: m.kahlmeyer@uni-kassel.de

N. Ratsch
e-mail: n.ratsch@uni-kassel.de

S. Böhm
e-mail: s.boehm@uni-kassel.de

L. F. M. da Silva and R. D. Adams (eds.), *6th International Conference on Adhesive Bonding 2021*, Proceedings in Engineering Mechanics,
https://doi.org/10.1007/978-3-030-87668-5_10

demanding applications due to pecularities and the lack of experience (Kohl et al. 2020).

In contrast to fossil and energy-intensive lightweight materials or reinforced plastics, wood is only reproducible, homogeneous and dimensionally stable to a limited extent, which impedes modelling, calculation and design. In addition, it is subject to conventional lightweight materials due to its lower transverse strength and rigidity as well as other technological boundary conditions. Factors such as inhomogeneity or anisotropy resulting from the wood morphology pose major challenges for its increased use.

The production of high-quality hybrid plywood composites is one way to counteract these deficits. This is often achieved by cleverly combining wood and other materials. The potential for lightweight construction is demonstrated in various publications (Kohl et al. 2013, 2016, 2017; Liu and Guan 2019; Lakreb et al. 2018; Dungani et al. 2019). Kohl et al. (2020) described the potential of hybrid wood-based materials for technical application such as automotive engineering; for this purpose, plywood boards made of beech veneer were reinforced by fibreglass mats, stainless steel foil or polymer films were produced and characterised (Kohl et al. 2020). This hybridisation results in an improvement of the bending strength and an increase in flexural modulus of elasticity. Additionally, the reinforced materials show a better dynamic fracture behaviour. Composites of wood and metals are promising hybrid materials, combining the stiffness, strength and fracture behaviour of metal with the low density of wood-based materials. However, adhesion between the composite layers is often challenging. In order to join such different materials, a variety of different adhesive types such as phenolic resorcinol, epoxy and polyurethane (PUR) adhesives lead to aging-resistant connections. But compared to aqueous dispersion adhesives they significantly reduce the sustainability of wood composites due to their high global warming potential.

For this reason this study aims at the development and evaluation of high performant hybrid wooden sandwich panels based on plywood combined with metals by sustainable adhesives. It was examined whether the hybrid material combination could be realised by an environmental friendly, water-based polyvinyl acetate (PVAc) so that the composite meets the high demands placed on technical applications. A PUR adhesive is used as a reference for this study.

2 Material and Methods

Tensile shear test specimens were carried out to examine whether different surface pre-treatments improve the adhesion and durability of PUR or PVAc bonded hybrid joints between wood and metal substrates.

Subsequently, sandwich panels of beech veneer and aluminium sheets were produced in a one-step pressing process to investigate bending properties and their durability.

2.1 Material

Tensile shear strength was determined by overlapping specimens made of 3 layered beech plywood (thickness 3.6 mm) from European Beech (*Fagus sylvatica*) and aluminium alloy ($AlMg_3$, thickness 1.6 mm). Two types of commercially available adhesives were used: 1-component PUR as well as PVAc (D3) from Jowat (Germany).

For the production of the sandwich panels, aluminium sheets with a thickness of 0.8 mm were used in addition to rotary cut beech veneers with a thickness of 1.2 mm.

2.2 Methods

2.2.1 Surface Treatment

Two different surface pre-treatments were used to investigate the influence of the surface conditions on the adhesive adhesion as well as on the mechanical properties of the sandwich composites. The focus was placed on the pre-treatment of the aluminium substrate, since prestudies led to delaminations in this interphase. Here, the physical processes of laser and plasma were selected for the surface treatment.

For the pre-treatment of the metallic substrates, the CL50 laser from Clean-Lasersysteme (Germany) was used with a pulsed, Q-switched Nd:Yag laser source with a wavelength: 1064 ± 4 nm in a workcell. The use of laser treatment enables the removal of oxidative residues and leaves a homogeneously prepared surface for the bonding process.

For surface pre-treatment with plasma the AS400 atmospheric pressure plasma system from Plasmatreat (Germany) was used. The distance between the nozzle and the surface was 10 mm. The maximum feed rate was 15 m/s in a meandering pattern.

In the experiment, the aluminium material was given the following surface pre-treatment: A: cleaning with isopropanol; B: plasma treatment (PT); C: laser treatment (LT); D: combination of laser and plasma treatment (LT + PT).

2.2.2 Single Lap Joints

Beech plywood with a density of 682 ± 37 kg/m^3 and aluminium test specimens were prepared for single lap joints. The overlap was 25 × 10 mm^2. The bonding area on the wood substrate was lightly roughened with sandpaper and then cleaned of dust before bonding. The metallic joining partner was cleaned with isopropanol and treated with by PT, LT or a combination of LT + PT. PUR or PVAc adhesive was then applied to the wood surface at an application rate of 200 g/m^2. Wood and metal substrates were joined with a pressure of 1 N/mm^2 and cured at room temperature according to the data sheet or after oven curing at 60 °C for one hour. After adhesive curing, the test specimens were conditioned under normal climate (20 °C/65% RH).

2.2.3 Manufacturing of Sandwich Composites

For manufacturing of hybrid wood sandwich composites (SC) beech veneers with a thickness of 1.2 mm and dimensions of 300 × 300 mm were combined with two aluminium alloy sheets (0.6 mm) as face materials. In a first step, adhesive (PUR or PVAc; 200 g/m^2) was applied on the veneers or metal surfaces and the layer structure was arranged – locked layer orientation of veneer: 0°/90°. The material stacks were placed in a laboratory up stroke hot press and compressed (constant pressure 1 N/mm^2) for 20 min at 80 °C. After panel manufacturing, the SC were removed from the press and cooled to room temperature at 20 °C and 65% RH.

2.2.4 Testing

Determination of tensile shear strength according to DIN 1465 and testing of sandwich panels for bending strength (DIN 53293) were used to characterise the bondings and the influence of the surface treatments on the strengths. In order to investigate the ageing behaviour of the adhesive systems and the hybrid joints or the SC, respectively, a climate change test was performed. Specimens for the determination of tensile shear strength and 4-point bending strength were aged for 20 cycles (10 days) and subsequently stored for 24 h at standard climate (20 °C; 65% RH).

2.2.5 Statistical Analysis

Shear strength and mechanical properties were statistically analysed by one way analysis of variance with the statistic software Origin 8.5. The SHAPIRO–WILK test and the LEVEN's test were applied to evaluate the normal distribution and homogeneity of variance. After confirmation, a TUKEY test was applied to find significant differences between the mean values. The statistical level was set to 5%.

3 Results and Discussion

3.1 Shear Strength of Single Lap Joints

Determination of shear strength by single lap joints according to EN 1465 allows the evaluation of the bonding qualities of joints as a function of adhesive and surface pre-treatment, see Fig. 1.

The results clearly show that surface treatment (B-D) significantly improves the strength. While samples that were only cleaned by isopropanol show average tensile shear strengths of <5 N/mm^2 (A), a significant improvement can be achieved through

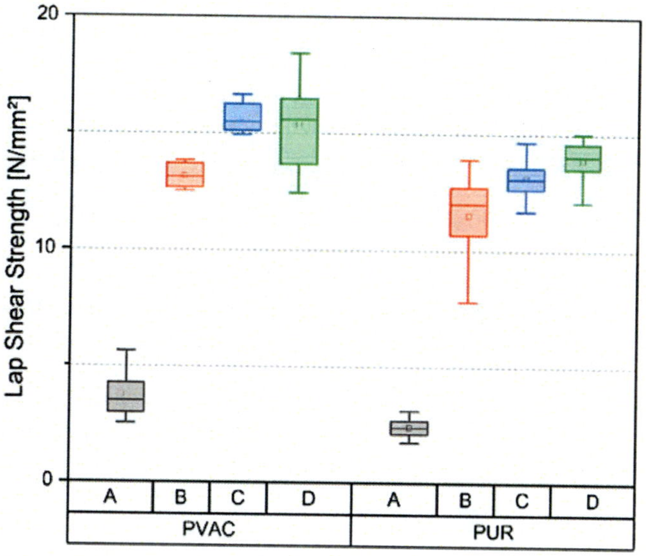

Fig. 1 Tensile shear strength of PUR and PVAc bonded wood-aluminium specimens; curing according to data sheet

physical pre-treatment. By the use of PT, LT or LT + PT, an enhancement to app. 12–15 N/mm^2 was possible until wooden fracture occurred (Figs. 1 and 2). These observations were accompanied by studies of the contact angle and surface free energy (SFE). As already described in literature, PT of aluminium increased the SFE and the polar components in the surface structure and can thus have a positive influence on wood-metal bonding using PVAc or PUR (Fig. 3) (Mui et al. 2017; Žigon et al. 2020).

An increase in SFE from the untreated state of 30.3 NM/m to the plasma-treated surface of 56.7 NM/m was obtained. This was also accompanied by an increase in the polar component from 5.6 to 25.6 NM/m.

Furthermore, the results show that an increase in temperature during adhesive curing improves the tensile shear strength. For specimens, which were bonded by PVAc (pre-treatment A), the curing at 60 °C increased the strength by a factor of 2 to approx. 7.7 N/mm^2. For PUR-samples (pre-treatment A) the thermal curing led to an increase by a factor of 4 to approx. 9.2 N/mm^2 (Table 1). This observation is attributed to the changed crosslinking of the adhesive.

A 10-day ageing period of the specimens causes a reduction in tensile shear strengths of about 35–50% for PVAc bonds and 16–20% for PUR specimens, although plasma or laser treated specimens do not fall to the level of untreated specimens.

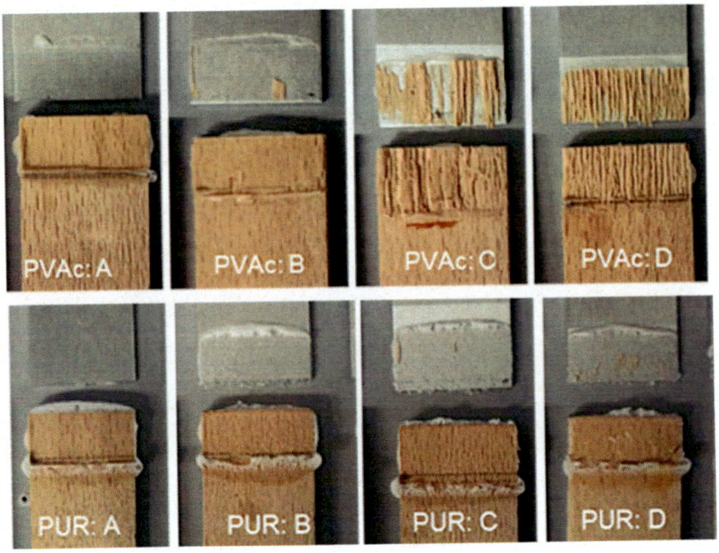

Fig. 2 Fracture pattern of aluminium-wood joints; curing according to data sheet

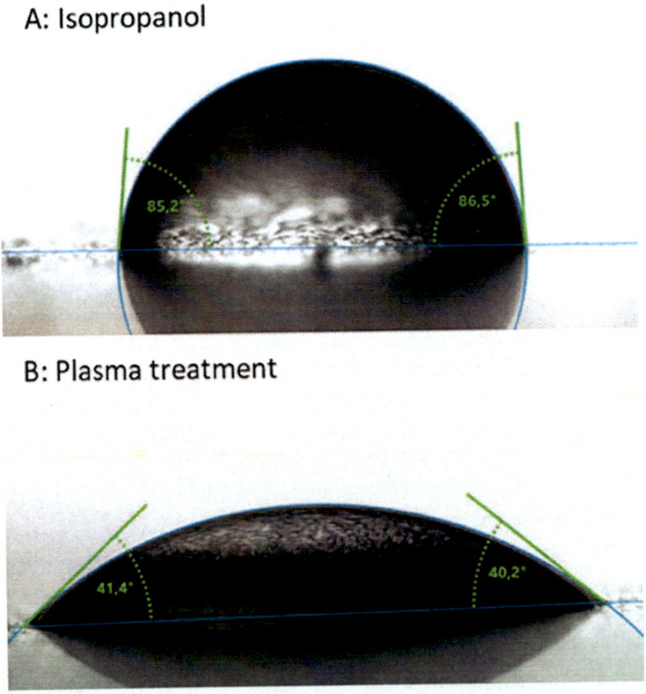

Fig. 3 Water droplets applied on untreated and plasma treated aluminium

Table 1 Determination of tensile shear strength (N/mm^2) of metal-wood joints after conditioning in normal climate and after 10 day ageing in climate change chamber; curing according to data sheet and accelerated at 60 °C in oven

		Testing after conditioning (20 °C/65% RH)		Testing after ageing	
Adhesive	Treatment	Curing: data sheet	Curing: accelerated	Curing: data sheet	Curing: accelerated
PVAc	A	3.73 (±1.10)	7.67 (±2.02)	3.44 (±0.71)	2.91 (±0.46)
	B	13.15 (±0.52)	14.92 (±1.51)	7.91 (±1.01)	2.99 (±0.28)
	C	15.66 (±0.67)	15.77 (±0.29)	6.92 (±0.70)	5.64 (±0.63)
	D	15.37 (±2.12)	14.10 (±1.51)	9.71 (±1.22)	5.66 (±0.62)
PUR	A	2.39 (±0.48)	9.20 (±1.07)	3.17 (±0.86)	5.05 (±1.42)
	B	11.51 (±2.11)	12.70 (±0.95)	9.91 (±1.41)	11.9 (±1.62)
	C	13.11 (±1.02)	14.29 (±0.93)	11.28 (±1.26)	15.16 (±1.21)
	D	13.88 (±1.06)	15.58 (±2.51)	11.97 (±0.47)	15.5 (±0.86)

3.2 *4-Point Bending Strength*

Figure 4 demonstrates the achieved bending strength at failure measured by a four-point bending test according to DIN 53293. Sample dimensions were 50 mm width and 180 mm length with a panel thickness of 8.1 mm; load and support length were 80 and 160 mm. The tests show that PUR-bonded sandwich panels achieve good strength values. During the bending test, a shear failure always occurs between the support and the load application point (Fig. 4). This means the wooden composite core fails due to shear stress and appears to be the weakest link in the composite.

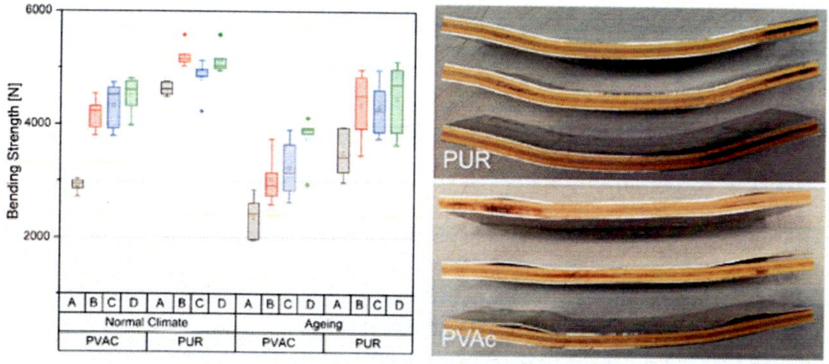

Fig. 4 *Left*: Bending strength of PUR or PVAc bonded sandwich panels; tested after conditioning phase in normal climate or after ageing test; *Right*: Fracture patterns of PUR or PVAc bonded sandwich panels after 4-point bending test

Regardless of any surface treatment of the aluminium, the sandwich panels show good mechanical properties.

In comparison, the PVAc-bonded sandwich panels show slightly lower strength values; it is clear from Fig. 4 that surface treatment results in an improvement of the flexural properties. If the aluminium sheets of the PVAc bonded sandwich panels were only cleaned with isopropanol, then an early shear fracture occurred between the aluminium and the core layer of plywood, but if the aluminium sheets were treated with PT, LT or a combination of both, flexural strength could be increased by 40%.

The sandwich panels were also subjected to a climate change test, and in the case of the PUR as well as PVAc bonded sandwich panels, it can be seen that high temperatures and humidity changes have a strong negative effect on the panel properties and on the bonding strengths.

However, it is clear from Fig. 4 that the use of surface treatments reduces the negative effect on strengths. Bending strength of untreated sandwich panels (A: Isopropanol) were reduced by approx. 18% (PVAc) or 24% (PUR), and strength of sandwich panels treated with PT or LT was reduced by 16% (PVAC:PT + LT) or 10% (PUR: LT) after ageing.

4 Conclusion

In the present study, investigations on the behaviour of hybrid wood-metal joints under tensile and bending loads as a function of various climates were carried out. In addition to the mechanical properties, the influence of surface treatments on the adhesion of the adhesives was examined.

The conclusions of this study are as follows:

- Tensile shear strength was improved by surface treatment, regardless of the adhesive choice (PUR or PVAc)
- Environmentally friendly plasma treatment of aluminium sheets made water-based dispersion adhesives suitable for stable wood-aluminium sandwich composite
- Ageing test of 10 days significantly reduced tensile shear strength of the specimens, but surface treatment prevented a drop to the level of untreated specimens
- PUR bonded sandwich panels, which were not exposed to ageing, did not require pre-treatment. When the composites were subjected to mechanical loading, shear failure always occurred. However, exposed to moisture and temperature, plasma or laser treatment still allowed stable joints
- For PVAc bonded sandwich panels a surface treatment of the aluminium was mandatory.

Acknowledgements The authors would like to thank the accompanying committee for their excellent support.

References

Dungani, R., Karliati, T., Hadiyane, A., Suheri, A., Suhaya, Y.: Coconut fibres and laminates with Jabon trunk (Anthocephalus cadamba Miq.) veneer for hybrid plywood composites: dimensional stability and mechanical properties. Eur. J. Wood Prod. (2019). https://doi.org/10.1007/s00107-019-01432-9

Kohl, D., Million, M., Böhm, S.: Adhesive bonded wood-textile compounds as potentially new eco-friendly and sustainable hightech materials. Adhesion Soc. 27–29 (2013)

Kohl, D., Link, P., Böhm, S.: Wood as a technical material for structural vehicle components. Proc. CIRP (2016). https://doi.org/10.1016/j.procir.2016.01.133

Kohl, D., Long, T.H.N., Böhm, S.: Wood-based Multi-material systems for technical applications—compatibility of wood from emerging and developing countries. Proc. Manuf. (2017). https://doi.org/10.1016/j.promfg.2017.02.078

Kohl, D., von Boyneburgk, C., Feldmann, M., Heim, H.-P., Böhm, S.: Characterization of wood-based multi-material systems under dynamic impact stress. Wood Mat. Sci. Eng. (2020). https://doi.org/10.1080/17480272.2018.1501605

Lakreb, N., Knapic, S., Machado, J.S., Bezzazi, B., Pereira, H.: Properties of multilayered sandwich panels with an agglomerated cork core for interior applications in buildings. Eur. J. Wood Prod. (2018). https://doi.org/10.1007/s00107-017-1198-3

Liu, Y., Guan, M.: Selected physical, mechanical, and insulation properties of carbon fiber fabric-reinforced composite plywood for carriage floors. Eur. J. Wood Prod. (2019). https://doi.org/10.1007/s00107-019-01467-y

Mui, T.S.M., Silva, L.L.G., Prysiazhnyi, V., Kostov, K.G.: Surface modification of aluminium alloys by atmospheric pressure plasma treatments for enhancement of their adhesion properties. Surf. Coat. Technol. (2017). https://doi.org/10.1016/j.surfcoat.2016.08.024

Žigon, J., Kovač, J., Zaplotnik, R., Saražin, J., Šernek, M., Petrič, M., Dahle, S.: Enhancement of strength of adhesive bond between wood and metal using atmospheric plasma treatment. Cellulose (2020). https://doi.org/10.1007/s10570-020-03212-8

Multiaxial Fatigue Life Assessment of Adhesive Joints Based on the Concepts of Critical Planes: Stress-Based Approaches

O. M. R. Beugre, A. Akhavan-Safar, and L. F. M. da Silva

Abstract Experimental observations show that nucleation and initial fatigue crack growth occur in a critical plane, a material plane with the most severe fatigue damage caused by stresses and strains acting on it. Accordingly, models based on critical fatigue damage plans are appropriate for life estimation of materials specially for metals where their ability to estimate life is mainly between the factor of ±2 and ±3 compared to the experimental results. Despite the successful and extensive application of critical planes technique for life assessment of metals, these methods have been rarely considered for fatigue life prediction of bonded joints. The current study presents an evaluation of the fatigue life estimation capacity of different multiaxial fatigue models based on the critical planes technique and using stress components. Subroutines were developed and used to analyze all material planes and all stress components to find a crack initiation plane where the fatigue parameter is maximum. Different fatigue damage models, including Findley, McDiarmid, Sines, and Crossland, were evaluated and compared using experimental fatigue data obtained at different mode mixities using Arcan joints. Using the calibrated models, fatigue lives of joints subjected to mixed-mode and pure mode loading conditions were estimated. Good agreement between the predictions and the experimental observations were found for McDiarmid, Sines, and Crossland criteria. Despite the lower accuracy in life prediction of the Findley model, it is still a safe method to estimate the fatigue life of bonded joints for mode II and mixed mode.

Keywords Critical plane approach · Stress-based multiaxial fatigue analysis · Fatigue damage model · Adhesive joints · Fatigue life estimation

---

O. M. R. Beugre · A. Akhavan-Safar (✉)
Institute of Science and Innovation in Mechanical and Industrial Engineering (INEGI), Rua Dr. Roberto Frias, 4200-465 Porto, Portugal
e-mail: aAkhavan-Safar@inegi.up.pt

L. F. M. da Silva
Departamento de Engenharia Mecânica, Faculdade de Engenharia, Universidade Do Porto Rua Dr. Roberto Frias, 4200-465 Porto, Portugal
e-mail: lucas@fe.up.pt

L. F. M. da Silva and R. D. Adams (eds.), *6th International Conference on Adhesive Bonding 2021*, Proceedings in Engineering Mechanics,
https://doi.org/10.1007/978-3-030-87668-5_11

1 Introduction

Adhesive joints are extensively used in a wide range of applications in different advanced sectors such as automotive and aerospace industries, because of several advantages that they offer compared to conventional mechanical joining techniques such as creating a more uniform stress distribution that results in a higher fatigue durability (da Silva et al. 2018; Akhavan-Safar et al. 2017; Castro Sousa et al. 2021). During their service life, adhesively bonded components are commonly subjected to cyclic loading conditions. Several studies have dealt with the fatigue analysis of bonded structures (Monteiro et al. 2019; Castro Sousa et al. 2020a, b). However, due to the complexity of the stress state within the adhesive layer and diverse parameters affecting the fatigue durability of joints, this topic is still open and challenging.

Fatigue life can be divided into two main phases, known as crack initiation and crack propagation but the initiation life is significant for joints subjected to low load levels. Fatigue crack initiation models are employed to predict the service life of joints subjected to load level cyclic loading conditions. A crack initiation approach is a total life method, which is usually referred to as stress-life (S–N) approach where the deformations are mainly elastic and consequently the joints experience high cycle fatigue (Sheriff et al. 2017). It is also worth noting that in industrial applications, adhesively bonded components often experience multiaxial loads. Furthermore, even if the loading is uniaxial, the adhesive layer (in most of the bonded joints) experiences a multiaxial stress state. Thus, the development of a fatigue failure criterion that accounts for the effect of different stress conditions is crucial. Accordingly, most of the stress-life based models proposed in papers are multiaxial stress-based models. For example, Hong Tae et al. (2015) investigated the applicability of a structural stress method developed for spot weld joints to predict the fatigue life of adhesive joints in lap-shear (LS) and coach-peel (CP) specimen geometries. The damage parameter of this method is an equivalent stress component. They concluded that the structural stress method can predict the fatigue life of adhesive joints regardless of the specimen geometry. Beber et al. (2016) analyzed the influence of plasticity on the lifetime prediction of bonded joints using the stress life approach, and the maximum principal stress was chosen as the equivalent stress. The effective stress was determined as the maximum value of the stress within the critical distance. They observed that the critical distance had a greater influence on predictions for single lap adhesive joints. The critical distance can be defined as a specific point away from the singular point where the failure parameter should be measured and, these criteria were proposed by some researchers to overcome the stress singularity at the bonding ends and also the mesh sensitivity issue (Akhavan-Safar et al. 2017). The work of Castro Sousa et al. (2020a, b) is among the most recent works on estimating the fatigue life of adhesively bonded structures using multiaxial stress-based models. In their study, they proposed an effective stress as a combination of the von Mises and the hydrostatic stresses to evaluate the fatigue life of adhesively bonded joints subjected to different mode mixities. These key stress parameters are measured along the mid-plane of the bondline at the location where the maximum principal stress is maximum, which is considered as the critical distance. They concluded that their proposed fatigue model was able to precisely predict the fatigue life of Arcan adhesive joints subjected to mixed mode fatigue loading conditions. Another research is that of Çavdar et al. (2020), in which the authors presented a concept for stress-based fatigue life prediction of multiaxial stressed adhesively bonded joints with a hyperelastic material behaviour. The influence of stress multiaxiality on fatigue strength was described by them with an equivalent stress hypothesis which is a combination of the second invariant of the deviatoric part of the Cauchy stress with the first invariant of the Cauchy stress. They concluded that the method adopted was efficient for stress-based fatigue lifetime prediction of adhesively bonded joints with a hyperelastic material behaviour under multiaxial loading. Beber and Schneider (2020) investigated fatigue failure criteria for structural film adhesive joints considering multiaxiality, mean stress, and temperature effects. The maximum principal stress model and the linear Drucker-Prager stress model were among the fatigue criteria used to assess the fatigue testing results. They concluded that both criteria worked well on fitting the multiaxial fatigue behaviour of joints, but the second criterion performed slightly better. Castro Sousa et al. (2021), in their work, proposed a model based on the stress–life approach for the estimation of the fatigue life of single lap adhesive joints with different geometries.

Many studies have been conducted to understand and predict the fatigue behaviour of adhesive joints under multiaxial loads, typically based on stress interaction equations. However, a few studies are found on critical planes models. One of the rare papers where the use of the critical plane stress model is investigated is that of the Beber et al. (2020), in which Findley's critical plane is used to assess the fatigue life of butt, scarf, and thick adherend shear test (TAST) specimens bonded with a structural film adhesive. The authors concluded that the Findley's critical plane criterion works well on fitting the multiaxial fatigue behaviour of joints.

Due to the successful applications of the critical plane technique to metallic materials, the current study aims to perform a rigorous evaluation of commonly applied multiaxial fatigue critical plane criteria for fatigue life assessment of bonded joints. Different critical plane stress-based multiaxial models including Findley, McDiarmid, Sines and Crossland were evaluated and compared using fatigue loading tests using Arcan joints experimental results available in the literature.

2 Stress-Based Multiaxial Fatigue Models

2.1 Critical Plane Concepts

Based on the experimental results, nucleation and early fatigue crack growth occur on a specific material plane called critical plane. The critical plane is a plane where the fatigue damage caused by the stresses and strains acting on it is the most critical

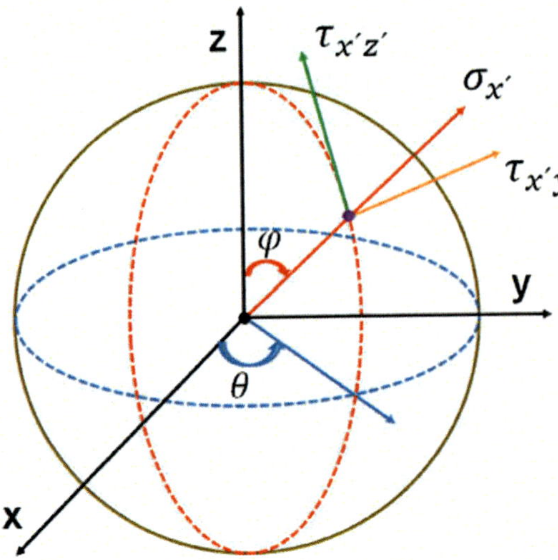

Fig. 1 The θ and φ angles defining local System $O_{x'y'z'}$ with respect to system O_{xyz}

(Stephens et al. 2001). To define the critical plane, a critical point should be determined first. For the homogenous materials, it is usually the point where the von Mises stress is maximum. After defining the maximum critical point, a sphere around this point should be formed as shown in Fig. 1. Depending on the model, specific stress and/or strain components should be analysed on all the planes around the critical point. The definition of the critical plane is different for different models and they usually present very different predictions for specific loading conditions. Therefore, it is important to check which model matches the predictions with the experimental data.

2.2 Stress-Based Critical Plane Fatigue Models

The stress–life methods are based on an assumption that stress controls the material behaviour at the local crack initiation site. Stress models are typically used for long life situations where stresses remain in the elastic zone. They are common methods of fatigue performance analysis of adhesive composite joints. The multiaxial fatigue analysis is used to estimate the fatigue life of a component under cycling loading when the crack is initiated due to a multiaxial state of stress. The assessment criteria for multiaxial fatigue loading can be based on three concepts: equivalent stresses, interaction equations, and critical plane approaches.

Findley (1957) introduced the first critical plane approach for stress-based multiaxial high-cycle fatigue and considered the plane of maximum shear stress as a critical damaging plane. According to this approach, a crack nucleation is governed by cyclic shear stress amplitude τ_a while maximum normal stress $\sigma_{n,max}$ on that plane assist in their early growth. Findley model is expressed as follows (Eq. 1):

$$\tau_a + k\sigma_{n,max} = f\left(N_f\right) \tag{1}$$

where k is a material constant, and the right-hand side of Eq. (1) is a function of the number of cycles to failure (N_f). The left-hand side of Eq. (1) can be determined by fitting the model to the stress-controlled torsional fatigue curve. Such a procedure yields the following expression (Eq. 2):

$$\tau_a + k\sigma_{n,max} = \tau_f'\left(2N_f\right)^{b_0} \tag{2}$$

The parameters τ_f' and b_0 are the shear fatigue strength coefficient and the exponent, respectively. The critical plane is the plane where $\tau_a + k\sigma_{n,max}$ is most extreme,$\left(\tau_a + k\sigma_{n,max}\right)_{max}$.

Taking the approach developed by Findley as a starting point, McDiarmid (1994) proposed the following expression for fatigue life estimation (Eq. 3):

$$\tau_a + \left(\frac{\tau_{-1}}{2\sigma_u}\right)\sigma_{n,max} = \tau_f'\left(2N_f\right)^{b_0} \tag{3}$$

where τ_{-1} is the fatigue limit under fully reversed torsion ($R = -1$) and σ_u is the ultimate strength of the material. But, in the McDiarmid approach, the critical plane is the plane where τ_a is the most extreme,$(\tau_a)_{max}$.

Sines (1959) developed his critical plane theory based on an equivalent stress state approach. This criterion states that the critical plane is the one where the octahedral shear stress amplitude ($\tau_{oct,a}$) achieves the maximum value, $\left(\tau_{oct,a}\right)_{max}$, which is responsible for the nucleation of crack, while the average hydrostatic stress for mean stresses ($\sigma_{h,mean}$) is responsible for their early growth. Such a fatigue model can be expressed as (Eq. 4):

$$\tau_{oct,a} + \left(\frac{3\tau_{-1}}{\sigma_0} - \sqrt{3}\right)\sigma_{h,mean} = \tau_f'\left(2N_f\right)^{b_0} \tag{4}$$

where σ_0 is fatigue limit under repeated tension ($R = 0$). $\tau_{oct,a}$ and $\sigma_{h,mean}$ are defined as follows (Eq. 5):

$$\tau_{oct,a} = \frac{1}{\sqrt{6}}\sqrt{\left(\sigma_{x,a} - \sigma_{y,a}\right)^2 + \left(\sigma_{y,a} - \sigma_{z,a}\right)^2 + \left(\sigma_{x,a} - \sigma_{z,a}\right)^2 + 6\left(\tau_{xy,a}^2 + \tau_{xz,a}^2 + \tau_{yz,a}^2\right)} \tag{5}$$

$$\sigma_{h,mean} = mean\left(\frac{\sigma_x + \sigma_y + \sigma_z}{3}\right) \tag{6}$$

where $\sigma_{x,a}$, $\sigma_{y,a}$ and $\sigma_{z,a}$ are the amplitudes of the normal stresses along the directions x, y, and z, respectively. $\tau_{xy,a}$, $\tau_{xz,a}$, and $\tau_{yz,a}$ are the amplitudes of total shear stresses. $\sigma_{x,mean}$,$\sigma_{y,mean}$, and $\sigma_{z,mean}$ are the normal mean stresses along the directions x, y, and z, respectively.

Crossland (1956) proposed a criterion very similar to that of Sines, disagreeing only on the influence of hydrostatic stress. For him, the maximum value of hydrostatic stress was the one who truly influenced fatigue life, not its mean value. Crossland then proposed his failure criterion as being (Eq. 7):

$$\tau_{oct,a} + \left(\frac{3\tau_{-1}}{\sigma_{-1}} - \sqrt{3}\right)\sigma_{h,max} = \tau_f^{'}\left(2N_f\right)^{b_0} \tag{7}$$

where σ_{-1} is the fatigue limit under fully reversed tension ($R = -1$), and $\sigma_{h,max}$ can be expressed as follows (Eq. 8):

$$\sigma_{h,max} = max\left(\frac{\sigma_x + \sigma_y + \sigma_z}{3}\right) \tag{8}$$

where σ_x, σ_y and σ_z are the normal stresses along the directions x, y, and z, respectively.

3 Description of the Experimental Program

Fatigue loading tests using Arcan joints, available in the literature (Castro Sousa et al. 2020a, b), were used to evaluate the critical plane models.

3.1 *Materials and Samples*

A toughened and high impact resistant two-component epoxy structural adhesive was considered in this study. This adhesive is designed for bonding of metallic surfaces namely steel and aluminium but can also be used to bond composite substrates (Castro Sousa et al. 2020a, b). Tensile properties of this adhesive are: Young's modulus E = 1159 MPa, Shear modulus G = 440 MPa and the Poisson's ratio υ = 0.32 (Castro Sousa et al. 2020a, b).

The Arcan joints shown in Fig. 2 have an overlap of 12.5 by 10 mm^2. The substrates were made of high strength steel (E = 210 GPa, υ = 0.3).

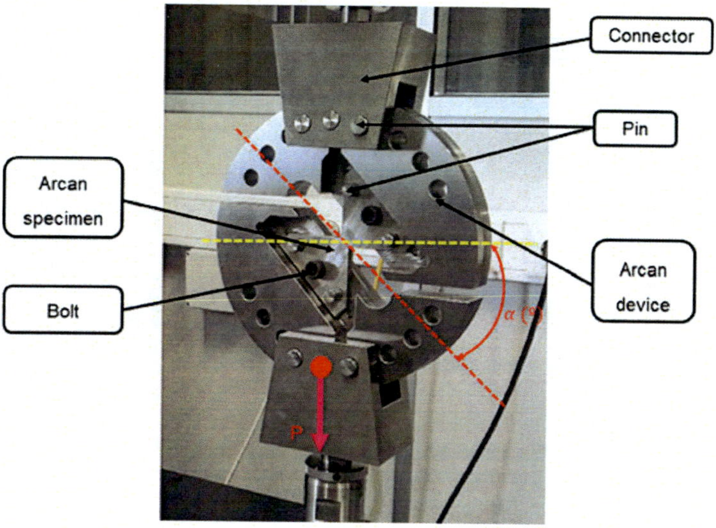

Fig. 2 Arcan apparatus and load angle definition

3.2 Test Device and Loading Conditions

All the experimental data were already conducted in a previous work (Castro Sousa et al. 2020a, b). All of the fatigue tests were carried under load control conditions with a sinusoidal waveform at a frequency of 10 Hz and the load ratio (R) of 0.1. The apparatus used for Arcan joints is shown in Fig. 2. Arcan joints were tested at three different loading angles including pure mode I (0°), mixed mode I + II (45°) and pure mode II (90°).

4 Fatigue Life Prediction Procedure

4.1 Finite Element Simulation

Stress analysis at the critical point is necessary to estimate the joint fatigue life using stress-based multiaxial fatigue models. Thus, a 2D linear elastic finite element analysis was conducted by using the standard commercial Abaqus code and was carried out at the maximum applied load applied during each fatigue test.

For the substrates, 3-node (triangular shaped) plain strain elements were used and the bondline were meshed using four-node plain strain elements. The mesh refinement was carried out near the free ends, as shown in Fig. 3. The loading and boundary conditions were applied to describe accurately the real conditions found

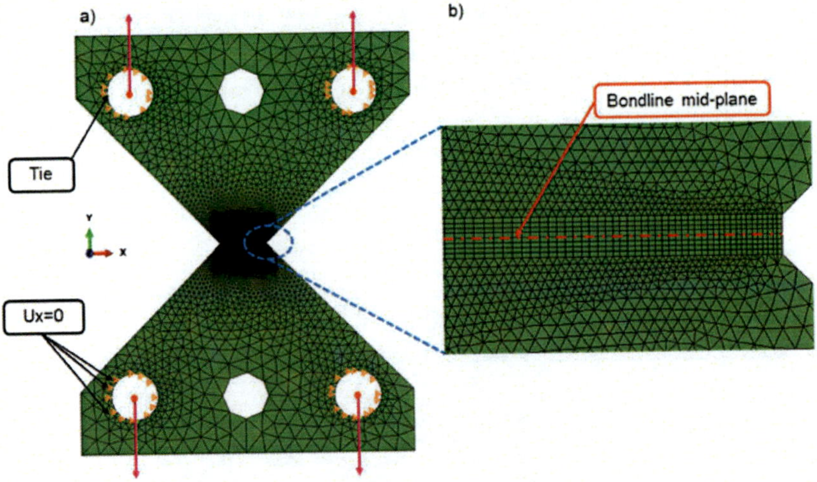

Fig. 3 **a** Mesh used for 2D simulation and boundary conditions for Mode I loading conditions, **b** Bondline mesh detail

on the experimental setup, as shown in Figs. 3a and 4. The load was applied to the substrate holes that are in contact with the pins and only the movement in the perpendicular direction to the applied load was restricted.

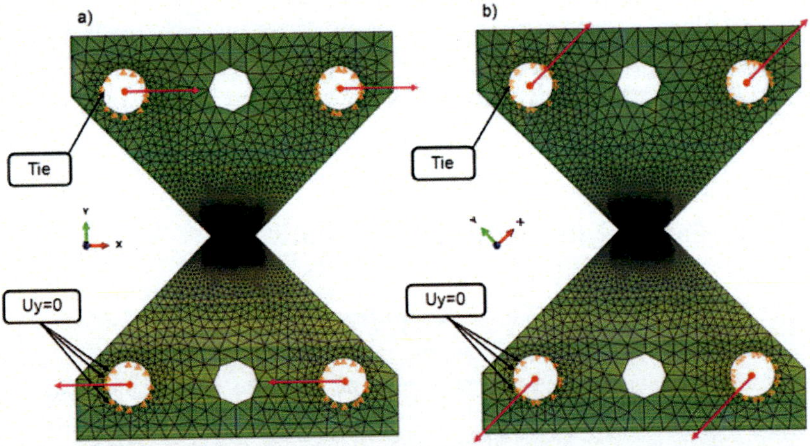

Fig. 4 **a** Boundary conditions for pure Mode II loading conditions, **b** mixed mode loading condition

4.2 Critical Plane Fatigue Models Constants

The parameters on the right side of fatigue model relations (see Eqs. 2, 3, 4, and 7) were obtained according to the following procedure: parameters $\tau_f^{'}$ and b_0 were determined by adjusting the Basquin curve (Eq. 4.1) to a set of experimental pure shear stress-life data (points circled in black in Fig. 5). As described in Dowling (2007) and Stephens et al. (2000), this adjustment was made by means of a linear regression of the data on a log–log scale and the independent variable adopted was the fatigue life (N_f).

$$\tau_a = \tau_f^{'}\left(2N_f\right)^{b_0}, \tag{9}$$

The next step is to obtain the value of k in the Findley model (hereafter denoted as $k_{Findley}$). In the current work, the same equivalent parameters were defined for other considered fatigue models where $k_{McDiarmid} = \frac{\tau_{-1}}{2\sigma_u}$, $k_{Sines} = \frac{3\tau_{-1}}{\sigma_0} - \sqrt{3}$ and $k_{Crossland} = \frac{3\tau_{-1}}{\sigma_{-1}} - \sqrt{3}$ adopted for the McDiarmid, Sines, and Crossland models, respectively.

For the Findley model, the value of k was first computed by fitting Eq. (2) to each mode I reference experimental results (shown in Fig. 5). Then the material constant

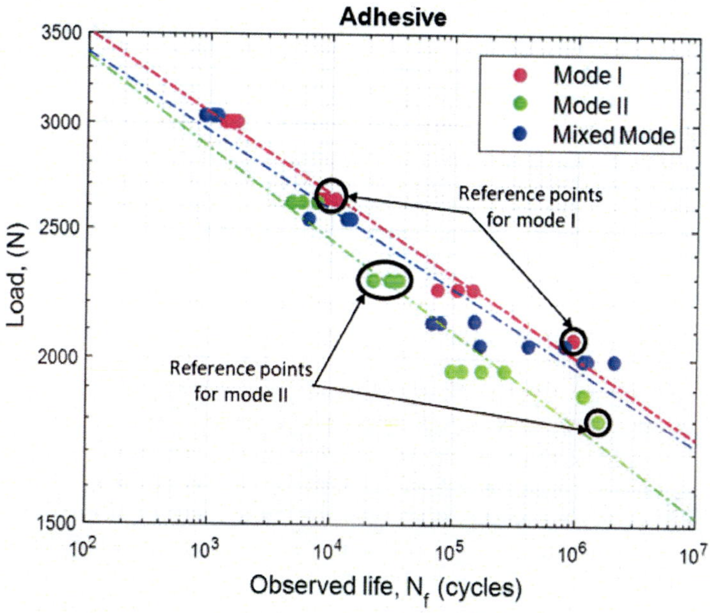

Fig. 5 Experimental data of Arcan joints for different mode mixities. Adapted from Castro Sousa et al. (2020a, b)

k for the Findley model (hereafter denoted as $k_{Findley}$) was obtained as the arithmetic mean value of the obtained k values. The same procedure was followed to calculated $k_{McDiarmid}$, k_{Sines}, and $k_{Crossland}$.

5 Results and Discussion

The fatigue test results of Arcan joints, for different mode mixities, are presented in Fig. 5. The vertical axis is the maximum applied load and the horizontal axis is the total fatigue life. The data covers different fatigue lives from 10^2 cycles to 10^7 cycles.

5.1 Finite Element Simulation Results

A finite element simulation was conducted to obtain the normal stresses (σ_{xx} and σ_{yy}) and the shear stress (τ_{xy}) at a critical location (critical point) in the initial plane which is the plane when $\theta = 0$ and $\varphi = 0$ (see Fig. 1). According to Castro Sousa et al. (2020a, b), for the Arcan joints, the normal stress and the shear stress should be measured along the adhesive mid-plane, shown in Fig. 3b, at the location where the maximum principal stress is higher, as shown in Figs. 7. Accordingly, the critical points for mode I and mode II are at 0.269 mm and 0.129 mm away from the bonding end, respectively (see Fig. 6). As illustrated in Fig. 7, the critical distance for the mixed mode is 0.329 mm. Using the obtained stresses for each load level and based on the fatigue models, fatigue lives were predicted.

5.2 Fatigue Models Evaluation

The obtained normal stresses (σ_{xx} and σ_{yy}) and the shear stress (τ_{xy}) at a critical location (Fig. 6) and in the initial plane (where θ and φ are zero) were used to calculate the stress components for each material plane defined by angles θ and φ (see Fig. 1), using strain and stress transformation rules.

As discussed above, the experimental data circled in black on Fig. 5 were used as reference points for calibration of the Findley, McDiarmid, Sines, and Crossland models. The calibrated models were used to estimate the fatigue life. Table 1 shows the obtained angles of the critical planes, according to the definition proposed by each model (Eqs. 2, 3, 4, and 7).

By knowing the fatigue parameter at each critical plan, the fatigue life can be estimated using Eqs. 2, 3, 4, and 7. However, before that, it is necessary to calculate the models' constants. Following the procedure discussed in Sect. 4, the constants of each model were obtained and listed in Table 2.

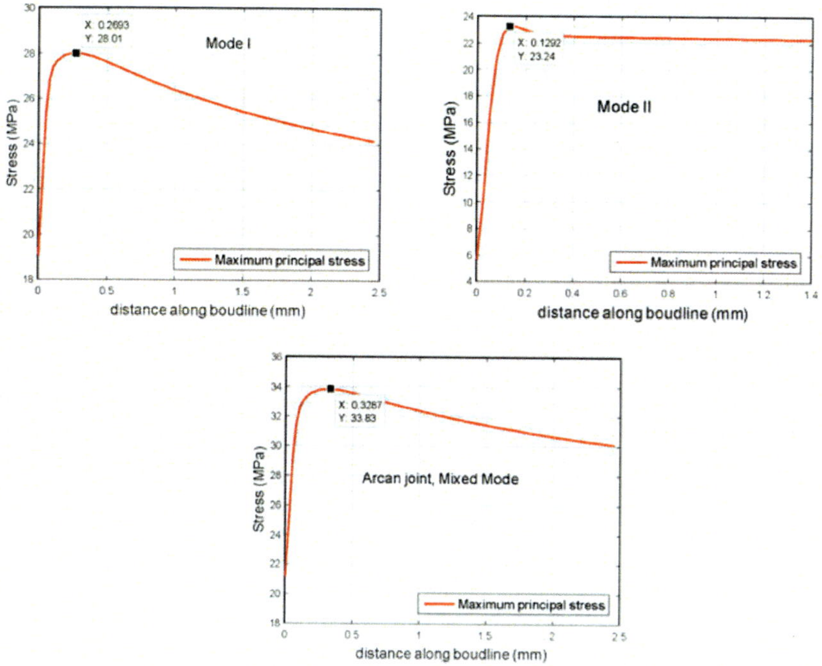

Fig. 6 Maximum principal stress and the critical distance for mode I, mode II and mixed mode

Figures 7, 8, 9, 10 compare the observed fatigue lives and the estimated fatigue lives using the four critical plane models introduced in this paper. In all figures, the central diagonal line represents a perfect correlation between the predicted lives and the observed lives, and the dashed lines represent factor 2 and factor 3 bandwidths. Life estimates below the central diagonal line are conservative and the ones above this line are non-conservative.

For the studied adhesive, the overall correlation produced by the Findley parameter is satisfactory. As shown in Figs. 7, 8, 9, 10 and 11, there is no mode II data within the bandwidths made by factor of 2, whereas most fatigue life estimations fall within the factor of 3 bandwidths compared to the experimental data. Except for mode I, using Findley criterion, predictions are conservative, that means that despite the lower accuracy in life prediction, Findley is still a safe model for fatigue life estimation of bonded joints. On the other hand, life estimation based on the McDiarmid criterion were very satisfactory for mode II loading conditions where the majority of the results are within the narrower bandwidths with the factor of 2. Using the McDiarmid model more than half of the data for the mode I and mixed mode are within the factor three band (see Fig. 5.8).

Sines could predict the fatigue life of the samples loaded in pure modes I and II. Fatigue life estimation results using Crossland show good correlation with the

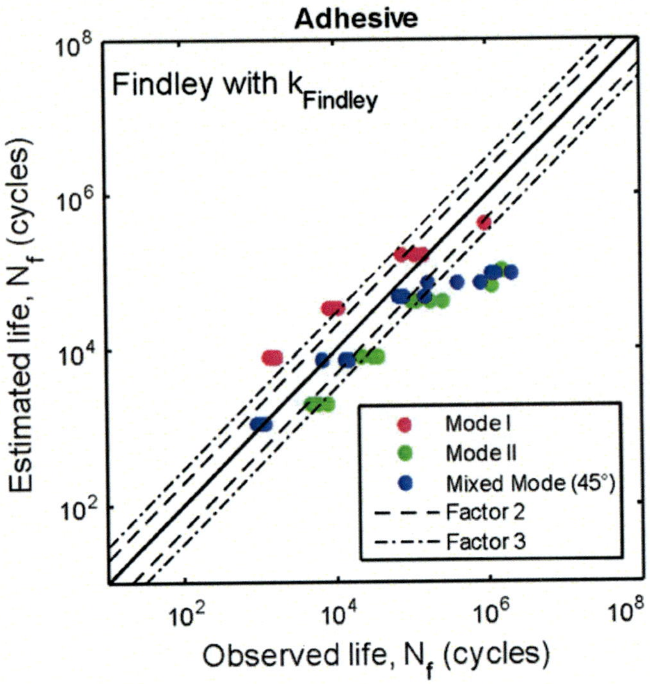

Fig. 7 Comparison between observed and estimated lives for the Findley parameter

Table 1 Angles of the critical planes

Model load	Findley		McDiarmid		Sines		Crossland	
	θ	φ	θ	φ	θ	φ	θ	φ
Mode I	88	57	88	45	64	48	64	48
Mode II	12	90	0	90	179	54	179	54
Mixed mode	68	57	68	45	58	2	58	2

Table 2 Cyclical properties of adhesive

τ'_f(MPa)	b_0	$k_{Findley}$	$k_{McDiarmid}$	k_{Sines}	$k_{Crossland}$
24.1791	−0.0953	0.195	0.25	0.5	0.245

experimental data. As shown in Figs. 5.6, 5.7 and 5.8, almost all of the estimated fatigue lives are within the bandwidths with the factor of 3.

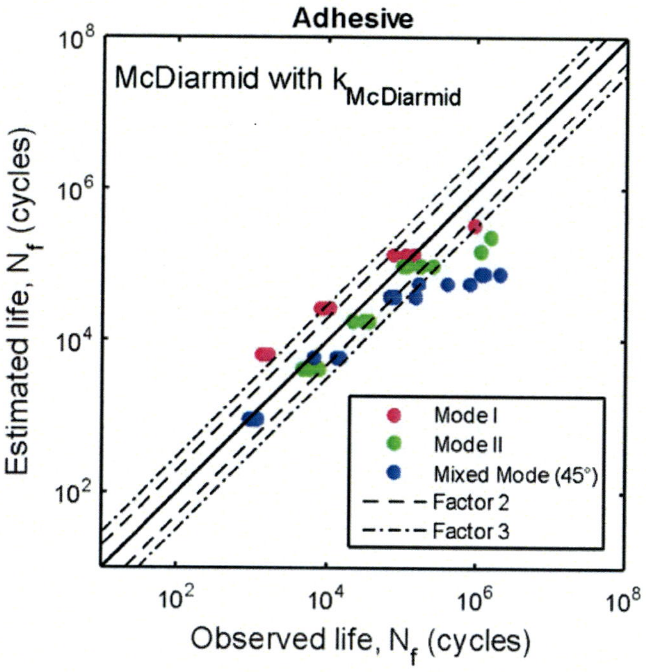

Fig. 8 Comparison between observed and estimated lives for the McDiarmid model

According to the results shown in Figs. 5.7 and 5.8, it was found that Sine is a good model for pure mode I fatigue life estimation of bonded joints while for pure mode II using the Crossland approach is recommended. For mixed mode all the same performance was observed for McDiarmid, Sines, and Crossland models.

6 Conclusions

The evaluation was based on Arcan joints loading tests. The fatigue testing results were assessed based on four critical plane fatigue criteria including Findley, McDiarmid, Sines, and Crossland. According to the results it was found that all the analyzed criteria worked well on fitting the multiaxial fatigue behaviour of Arcan joints tested in pure modes I and II fatigue loading except for the Findley criterion which is less accurate than the other criteria in pure mode II loading conditions. The less accuracy

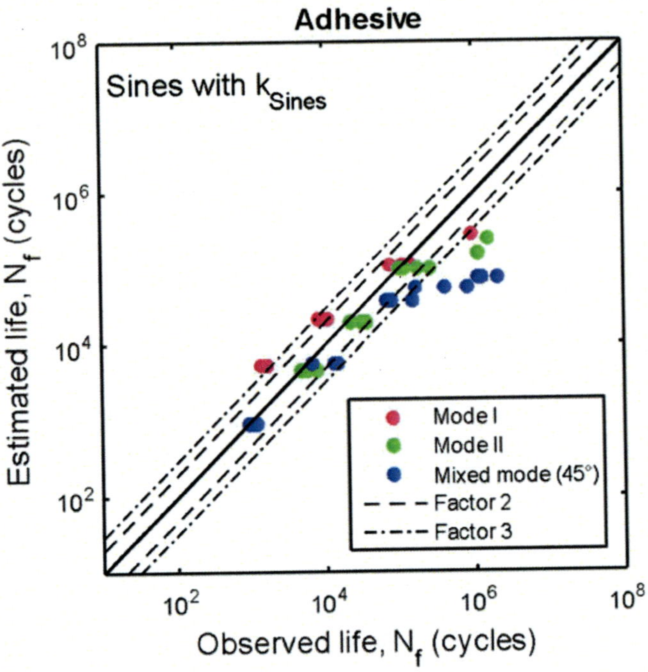

Fig. 9 Comparison between observed and estimated lives for the Sines parameter with k_{Sines}

of the model in pure mode I compared to mode II fatigue results is due to the fact the criteria used in the current paper are shear based fatigue models and therefore they are not the right models for tensile loading conditions. However, since in real practice the joints are mainly subjected to shear loading, these models should be good choices for real joints. For mixed mode conditions, the overall correlation produced by the four analyzed models was in good agreement with the experimental data. Half of the estimated lives are within the narrow bandwidth with the factor of 2 and the majority of the estimations are within the bands of 3. It should be noted that all the predictions outside of the bands are conservative where the estimated life is lower than the experimental observations which is safe for design purposes. Finally, according to the results it can be concluded that Sine is a good model for pure mode I while for pure mode II the Crossland approach is recommended. For mixed mode loading conditions, predictions of McDiarmid, Sines, and Crossland models are in good agreement with the test data.

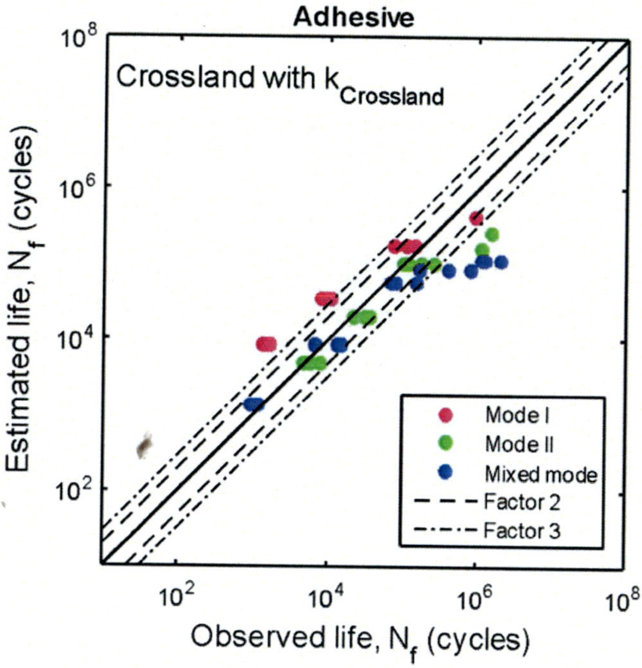

Fig. 10 Comparison between observed and estimated lives for the Crossland parameter with $k_{Crossland}$

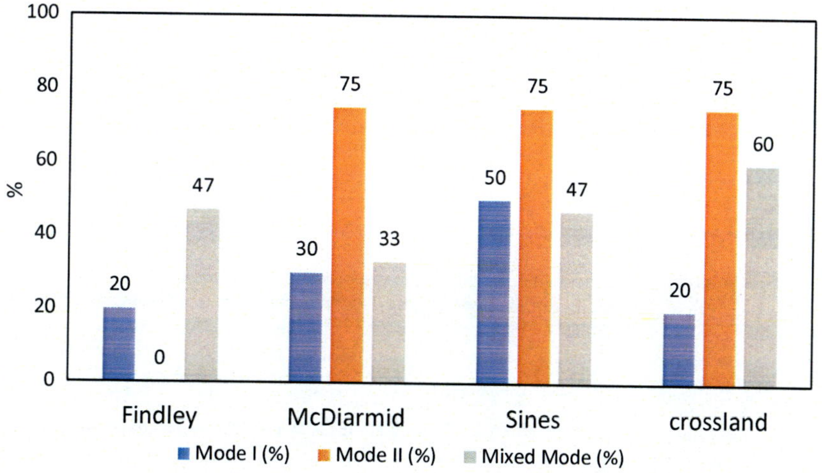

Fig. 11 Percentages of estimated lives within the bandwidths of factor of 2

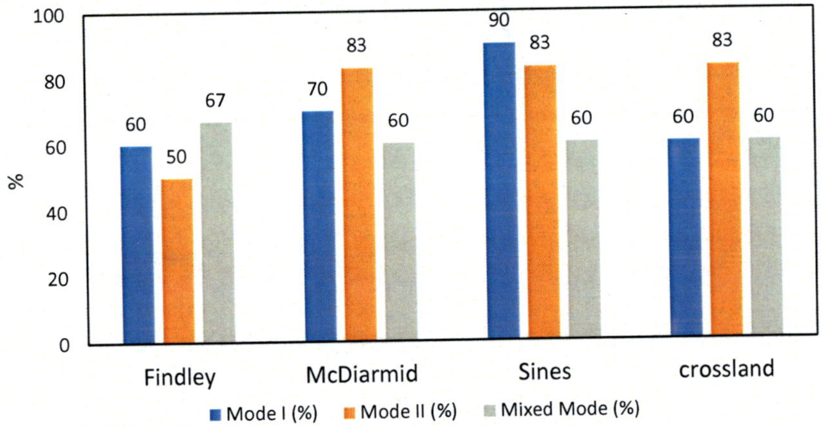

Fig. 12 Percentages of estimated lives within the bandwidths of factor of 3

References

Akhavan-Safar, A., da Silva L.F.M., Ayatollahi M.R.: An investigation on the strength of single lap adhesive joints with a wide range of materials and dimensions using a critical distance approach. Int. J. Adhes. Adhes. 78248–255. https://doi.org/10.1016/j.ijadhadh.2017.08.009 (2017)

Beber, V.C., Fernandes, P.H.E., Fragato, J.E., Schneider, B., Brede, M.: Influence of plasticity on the fatigue lifetime prediction of adhesively bonded joints using the stress-life approach. Appl. Adhes. Sci. **4**(1), 1–18 (2016)

Beber, V.C., Schneider B.: Fatigue of structural adhesives under stress concentrations: notch effect on fatigue strength crack initiation and damage evolution. Int. J. Fatigue 140105824 (2020) https://doi.org/10.1016/j.ijfatigue.2020.105824

Castro, F., Akhavan-Safar, S.A., Goyal, R., da Silva L.F.M.: Fatigue life estimation of single lap adhesive joints using a critical distance criterion: an equivalent notch approach. Mech. Mater. (2021) 153103670. https://doi.org/10.1016/j.mechmat.2020.103670

Castro Sousa, F., Akhavan-Safar, A., Goyal, R., da Silva, L.: The influence of mode mixity and adhesive system on the fatigue life of adhesive joints. Fatig. Fract. Eng. Mater. Struct. **43**(10), 2337–2348 (2020a)

Castro Sousa, F., Akhavan-Safar, A., Goyal, R., da Silva, L.F.M.: Fatigue life estimation of adhesive joints at different mode mixities. J. Adhes: 1–23 (2020b)

Crossland, B.: Effect of large hydrostatic pressures on the torsional fatigue strength of an alloy steel. In: Proceedings of the International Conference on Fatigue of Metals ImechE London, UK, pp. 138–149 (1956)

Dowling, N.E.: Mechanical Behavior of Materials, 3rd edn. Pearson Education, Inc. (2007)

Findley, W.N.: Fatigue of metals under combinations of stresses. Trans. A.S.M.E **79**, 1337–1348 (1957)

Gonçalves, V.M., Martins, P.A.F.: Joining stainless steel parts by means of weld bonding. Int. J. Mech. Mater. Des. **3**, 91–101 (2006)

HongTae, K., Zhen, L., Khosrovanehb, A.K., Byeongwook, K., Lia, Z.: Fatigue life predictions of adhesive joint of sheet steels. 6th Fatigue Des. Conf. Fatigue Des. Proc. Eng. **133**, 518–527 (2015)

McDiarmid, D.: A shear stress based critical-plane criterion for multiaxial fatigue failure for design and life prediction. Fatigue Fract. Eng. Mater. Struct. **17**(12), 1475–1485 (1994)

Monteiro, J., Akhavan-Safar, A., Carbas, R., Marques, E., Goyal, R., El-zein, M.: Mode II modeling of adhesive materials degraded by fatigue loading using cohesive zone elements. Theor. Appl. Fract. Mech. **103**, 1–7 (2019)

Olajide, S.O., Arhatari, B.D.: Progress on interacting fatigue creep & hysteretic heating in polymer adhesively bonded composite joints. Int. J. Fatigue 9868-80. https://doi.org/10.1016/j.ijfatigue.2017.01.021 (2017)

da Silva, L.F.M., Öchsner, A., Adams, R.: Handbook of Adhesion Technology. Springer Nature, Cham (2018)

Sines, G.: Behavior of metals under complex static and alternating stresses. In: Metal Fatigue, pp. 145–169. McGraw-Hill, New York (1959)

Stephens, R.I., Fatemi, A., Stephens, R.R., Fuchs, H.O.: Metal Fatigue in Engineering. John Wiley, New York (2001)